Gene Expression and Phenotypic Traits

Edited by Yuan-Chuan Chen and Shiu-Jau Chen

Published in London, United Kingdom

IntechOpen

Supporting open minds since 2005

Gene Expression and Phenotypic Traits
http://dx.doi.org/10.5772/intechopen.82955
Edited by Yuan-Chuan Chen and Shiu-Jau Chen

Contributors
Aleksander Smirnov, Antonina Trukhina, Ilke Onen, Ebru Alp, Atiye Seda Yar Saglam, Chanda Siddoo-Atwal, Jade Q. Clement, Michihiko Ito, Morgan Salmon, Sergey Baranov, Yuan-Chuan Chen, Azza El-Demerdash, Neveen Bakery

First published in London, United Kingdom, 2020 by IntechOpen
IntechOpen is the global imprint of INTECHOPEN LIMITED, registered in England and Wales, registration number: 11086078, 7th floor, 10 Lower Thames Street, London, EC3R 6AF, United Kingdom
Printed in Croatia

British Library Cataloguing-in-Publication Data
A catalogue record for this book is available from the British Library

Additional hard and PDF copies can be obtained from orders@intechopen.com

Gene Expression and Phenotypic Traits
Edited by Yuan-Chuan Chen and Shiu-Jau Chen
p. cm.
Print ISBN 978-1-83880-031-4
Online ISBN 978-1-83880-032-1
eBook (PDF) ISBN 978-1-83880-318-6

Meet the editors

Yuan-Chuan Chen obtained a PhD in Biochemistry at the University of California, Berkeley, USA, in 2015. His research interests include pharmacy/pharmacology, biochemistry, microbiology/virology, cell/molecule biology, biotechnology/nanotechnology, cell/gene therapy and policy/regulation. His studies focus on the discovery, application, perspectives and challenges of biopharmaceuticals and CRISPR/Cas9 technologies. Additionally, he is interested in basic research, multiple applications and therapeutics about gene expression and phenotypic traits.

Shiu-Jau Chen obtained an MD and PhD in Anatomy and Cell Biology at the National Taiwan University in 1994 and 2013, respectively. His specialties are neurosurgery and brain disease treatment. His studies focus on the prevention and treatment of drug addiction and neurodegenerative diseases. He is also interested in the treatment of diseases associated with gene expression and phenotypic traits.

Contents

Preface

Gene expression is a process by which genetic information is used in the synthesis of functional products including proteins and functional RNAs, such as transfer RNA (tRNA), small nuclear RNA (sn RNA), micro RNA (miRNA), small/short interfering RNA (siRNA), among others. This process is utilized by eukaryotes, prokaryotes, and viruses to generate macromolecules for constituting cellular components and exhibiting living functions. Several steps in the gene expression process could be regulated, such as transcription, post-transcriptional modification, translation, and post-translational modification. Regulation of gene expression is to modulate the production amount and timing of the functional products. Control of gene expression is critical to allow cells to produce their functional products when cells need them; in turn, this gives cells the flexibility to adapt to a variable environment, and respond to external signals, stimuli, and damages. Cellular structures and functions can also be controlled by gene expression regulation. Consequently, the regulation is crucial for cells to proliferate, differentiate, transport, metabolize, and repair. It is advantageous to the versatility, development, and adaptability of living organisms. A phenotypic trait, the expression of genes in an observable way, is an obvious and measurable trait. For example, hair color is a phenotypic trait and genotype composed of genes that determine hair color, but the hair color observed is the phenotype. The phenotype is variable depending on the genetic make-up of the organism, and influenced by the surroundings to which the organism is subjected across its morphogenesis, including various epigenetic processes. In this book, these findings and their implications are broadly discussed, and future genomic and phenotypic studies, analyses, and applications are highlighted. In this book, we explore the significance, mechanism, function, characteristic, determination and application of gene expression and phenotypic traits.

We are grateful for the participation of all authors and IntechOpen's enthusiasm in helping us to complete and publish this book.

Yuan-Chuan Chen, PhD
Program in Comparative Biochemistry,
University of California, Berkeley,
CA, USA

Shiu-Jau Chen, MD, PhD
Department of Neurosurgery,
Mackay Memorial Hospital,
Taipei, Taiwan

Section 1

Background

Chapter 1

Introductory Chapter: Gene Expression and Phenotypic Traits

Yuan-Chuan Chen

1. Gene expression

Gene expression is a process by which the genetic information is used in the synthesis of functional products including proteins and functional RNAs (e.g., tRNA, small nuclear RNA, microRNA, small/short interfering RNA, etc.). The process of gene expression is applied by all organisms including eukaryotes, prokaryotes, and viruses to produce the macromolecular machinery for life. Through controlling the cell structure and function, the gene plays an important role in cellular differentiation, morphogenesis, adaptability, and diversity. Because the control of the timing, location, and levels of gene expression can have a significant effect on gene functions in a single cell or a multicellular organism, gene regulation may also drive evolutionary change. Several steps in the gene expression process can be regulated, such as transcription, posttranscriptional modification (e.g., RNA splicing, 3′ poly A adding, 5′-capping), translation, and posttranslational modification (e.g., protein splicing, folding, and processing).

1.1 Transcription

The genomic DNA is composed of two antiparallel strands with 5′ and 3′ ends which are reverse and complementary for each. Regarding to a gene, the two DNA strands are classified as the "coding strand (sense strand)," which includes the DNA version of the RNA transcript sequence, and the "template strand (antisense strand, noncoding strand)" which serves as a blueprint for synthesizing an RNA strand. During transcription, the DNA template strand is read by an RNA polymerase to produce a complementary and antiparallel RNA primary transcript. Transcription is the first step of gene expression which involves copying a DNA sequence to make an RNA molecule including messenger RNA (mRNA), ribosome RNA (rRNA), and transfer RNA (tRNA) by the principle of complementary base pairing. In prokaryotes, transcription is performed by enzymes called RNA polymerases to form all RNA molecules. In eukaryotes, it is mainly performed by RNA polymerases I, II, III, IV, and V to make rRNA, mRNA, tRNA, etc. (**Table 1**).

1.2 Posttranscription modification

Posttranscriptional modification is referred to as biological processes by which RNA primary transcripts are chemically changed posterior to transcription in eukaryotes. This process significantly modifies the chemical structure of RNA molecules by three main constituting steps: the putting of a 5′ cap, the addition of a 3′ polyadenylated tail, and RNA splicing. In eukaryotes, primary RNA transcript must be processed after transcription because the initial precursor mRNA (primary RNA transcript) usually contains both exons (coding sequences) and introns (noncoding

RNA polymerase	Function
RNA polymerase I	Synthesis of precursor 45S rRNA (35S in yeast) which matures into 28S, 18S, and 5.8S rRNAs; the formation of the major RNA sections of the ribosome
RNA polymerase II	Synthesis of mRNA precursors; most snRNA and miRNA
RNA polymerase III	Synthesis of tRNAs, 5S rRNA and other small RNAs found in the nucleus and cytosol
RNA polymerase IV	Synthesis of siRNA in plants
RNA polymerase V	Synthesis of RNAs involved in siRNA-directed heterochromatin formation in plants

Abbreviation: mRNA, messenger RNA; rRNA, ribosome RNA; tRNA, transfer RNA; snRNA, small nuclear RNA; miRNA, microRNA; siRNA, small/short interfering RNA.

Table 1.
RNA polymerase function in eukaryotes.

sequences). By removing the introns to link the exons directly in RNA splicing, such processing is able to ensure the correct translation of eukaryotic genomes. Also, posttranscription modification can protect primary RNA transcripts from degradation in that the 5′ cap and poly-A tail are able to protect the transcripts and facilitate the transportation of the mRNAs to ribosomes.

1.3 Translation

Every mRNA is composed of a 5′ untranslated region (5′UTR), a protein-coding region or open reading frame (ORF), and a 3′ untranslated region (3′UTR). The ORF carries information for protein synthesis encoded by the RNA triplet codes. The codon, a RNA triplet of the coding region, corresponds to a binding site complementary to an anticodon triplet in tRNAs. tRNAs with different anticodon triplets carry different amino acids and thereby make amino acids link together according to the order of triplets in the coding region by the assistance of ribosomes.

In prokaryotes, translation usually occurs in the cytosol where the specific amino acid, enzymes, and small subunits of the ribosome bind to the tRNA. Translation simultaneously proceeds with transcription (co-transcriptionally), using a mRNA that is still being synthesized. In eukaryotes, transcription occurs in nucleus, and translation occurs in cytoplasm (not co-transcriptionally). Although translation can occur in a variety of regions of the eukaryotic cells, its major working locations are the cytoplasm for soluble cytoplasmic proteins and the endoplasmic reticulum (ER) membrane for secretory proteins or integral proteins.

1.4 Posttranslational modification

Posttranslational modification is referred as the chemical change which proteins may undergo after translation. The main alternations are the specific cleavage of precursor proteins, the excision of signal recognition peptides (SRP), the formation of disulfide bonds, the covalent addition or removal of low-molecular-weight groups, and the addition of metal ions, leading to protein modifications such as glycosylation, lipidation, hydroxylation, methylation, mono-ADP-ribosylation, myristoylation, oxidation, palmitoylation, and phosphorylation. It plays a crucial role in the regulation of the protein folding, the targeting of specific subcellular compartments, the interaction of ligands or other proteins, and the state of functions, such as enzyme catalytic activity or the signaling function of proteins in signal transduction pathways.

2. Reverse transcription

Reverse transcription is referred to as the synthesis of complementary DNA (cDNA) from an RNA template using reverse transcriptase (RTase). For reverse transcription, RT uses an RNA template and a short RNA primer complementary to the 3′ end of the RNA to direct a cDNA strand synthesis from 5′ end to 3′ end. The first-strand cDNA can be made to be double-stranded using DNA polymerase I, and ribonuclease (RNase) H activity for RNA primer digestion is required in this case. The engineered RTase can improve the efficiency of full-length cDNA products, ensure the completeness of the mRNA transcript copying, and enable the propagation of a faithful DNA copy of an RNA sequence. The use of the thermostable RTase is very helpful for the denaturation of RNA structure when the RNA sequence contains high amounts of secondary structure. Some retroviruses (e.g., human immunodeficiency virus, HIV) which have an RNA genome are able to transcribe RNA into cDNA using RTase. The RNase H digests the RNA strand, and then the cDNA strand is used as a template to synthesize a complementary DNA strand to form a double helix DNA structure. The resulting double stranded DNA (ds DNA) can be integrated into the host DNA genome, causing the host cell to produce viral proteins that assemble into new virions. In HIV, the host cell undergoes apoptosis (programmed cell death).

Telomerase (terminal transferase, RNA-directed DNA polymerase) is a kind of RTase that lengthens the ends of linear chromosomes in some eukaryotes, containing an RNA template from which it synthesizes a repeating sequence of DNA or "junk" DNA. It is active in stem cells, gametes. and most cancer cells to enable these cells to replicate their genomes immortally without losing important protein-coding DNA sequence. Embryonic stem cells highly express telomerase to allow these cells to divide repeatedly and form the individual such as male sperm cells, epidermal cells, and activated T and B lymphocytes. However, it is usually absent from or at very low levels in most somatic cells. The repeated DNA sequence (telomere) can be considered as a "cap" for a chromosome. In normal cells, the telomere is shortened when a linear chromosome is duplicated. Activation of telomerase is one of the processes that let cancer cells become indefinite. Telomerase allows each offspring to avoid losing a bit of DNA, making the normal cells divide without limitation and become abnormal cells, and the unbounded cell growth is a characteristic of cancer.

3. Epigenetics

Epigenetics is the study of heritable phenotype changes that do not involve alterations in the DNA sequence, meaning a change in phenotype without a change in genotype. The term also refers to the functionally associated changes to the genome that do not encompass a change in the nucleotide sequence. At present, DNA methylation, histone modification, and noncoding RNA (ncRNA)-associated gene silencing are major functions for involving in the initiation and support of epigenetic changes. It often indicates changes that affect gene activity and expression and shows phenotypic changes which can be transferred to the offspring.

Epigenetic changes can be influenced by several factors including the age, environment, lifestyle, and disease; it is traditionally considered to be regular and natural. This kind of change may be consistent through cell divisions during the cell life cycle and may also sustain for several generations, even if they do not have changes in the organism's underlying DNA sequence. The process of cellular differentiation is an example of an epigenetic change in eukaryotic biology.

3.1 The evolving landscape of epigenetic research

In 1969, Griffith and Mahler first suggested that DNA methylation might play a crucial role in long-term memory function. DNA methylation has currently become one of the most extensively studied and well-characterized epigenetic modifications. The other main modifications include chromatin remodeling, histone modifications, and noncoding RNA mechanisms. The new findings about epigenetics are the correlation between epigenetic changes and diseases such as cancers, mental retardation, immune disorders, neuropsychiatric disorders, and pediatric disorders.

3.2 Environment and lifestyle can influence epigenetic change from one generation to the next

Not only the environment but also individual lifestyles can directly interact with the genome to affect epigenetic changes. These epigenetic changes may be demonstrated at various stages throughout an individual's life and even last to his offspring. The prenatal and early postnatal environmental factors can influence the adult risk for the incidence of various chronic diseases and behavioral disorders in human epidemiology studies. It is known that the children have elevated rates of coronary heart disease and obesity after their mothers are exposed to famine during early pregnancy compared with those who are not exposed. Additionally, maternal exposure to air pollution could affect her children's asthma susceptibility. Similarly, the adults who were prenatally exposed to famine have also been reported to have significantly higher incidence of schizophrenia. Fortunately, maternal ingestion of vitamin D is capable to adjust DNA methylation that impacts placenta function.

3.3 Environment and lifestyle affect individual epigenetics and health

Epigenetics are considered to be dynamic and changeable by the influence of lifestyle options and environmental factors, though our epigenetic marks are more stable during adulthood. Epigenetic effects gradually occur both in the womb and the full course of a human life span, and epigenetic changes could be reversed. Epigenetics have shown that different lifestyle options and environmental exposures can change DNA marks and play a vital role in the determination of health outcomes. The environment can dominantly influence the epigenetic tags and disease susceptibility. Pollution has become a significant topic because scientists have found air pollution could induce DNA methylation and increase one's risk for neurodegenerative disease. Fortunately, vitamin B groups potentially protect humans from harmful epigenetic effects of pollution and against the other harmful effects on the body.

3.4 Deciphering the relationship between epigenetics and diseases

It is known that chronic pancreatitis causes a high risk of inflammation-associated progression to pancreatic cancer. The difficulty in rapidly diagnosing the disease is closely associated with its high mortality rate. Previous studies have demonstrated that cell-free DNA methylation from inflammatory diseases or cancer is variable, thereby opening a new era in developing biomarkers for the early diagnosis of diseases. Hence, early diagnosis for pancreatic cancer becomes crucial and facilitates the related studies into the epigenetic profiles [1]. Natale et al. reported that exploiting the relationship between abnormally methylated cell-free DNA and pre-neoplastic lesions or chronic pancreatitis may become a novel method in developing tools for the early diagnosis of pancreatic cancer. Early diagnosis potentially makes it possible for the prediction of prognosis, the monitoring of tumor progression, and the development of effectively therapeutic strategies and provides precision medicine for patients suffering from a pancreatic disease [1].

The early-life environment including air quality is known to be critical for fetal programming. The air pollution exposure to mothers during pregnancy may adversely influence newborn outcomes such as baby birth weight, preternatural birth, and preterm birth. Therefore, it is needed to understand both air pollution-induced early health effects and its later-life consequences. Saenen et al. provided an overview of air pollution-induced placental molecular changes observed in the ENVIRONAGE birth cohort and assess the existing evidence. They reported that nitrosative stress and epigenetic alterations in the placenta may result from the prenatal exposure to air pollution [2]. It is crucial to realize the clinical consequences of early-life epigenetic changes in the follow-up of child or birth cohort study. The public health policy maker should have understanding of epigenetic consequences and transgenerational risks to propose effective strategies which are focusing on providing effective protection of pregnant women, unborn children, and infants against exposure to adverse lifestyle factors [2].

4. Phenotypic traits

Genotype, a unique genome that can be revealed by genomic sequencing is a complete inheritable genetic identity. It is mediated by a special gene, cluster of genes, or set of genes which are carried by an individual. Genes are certain DNA segments that code for the protein production to determine distinct traits of individuals. DNA contains the genetic code which is responsible for all cellular functions such as mitosis, meiosis, DNA replication, protein synthesis, molecule transportation, etc. Each gene is located on a chromosome and can exist in different forms called alleles which are located on specific regions of chromosomes. The alleles can be transmitted from parents to offspring through sexual reproduction. The diploid organism inherits two alleles for each gene; one allele is from the father, and the other allele is from the mother. The interactions between alleles determine an organism's phenotype.

The phenotype is a description of actual physical characteristics and encompasses directly visible characteristics such as height, weight, skin color, eye color, size, shape, health condition, disease history, even behavior, and temperament. However, not all phenotypes are a direct result of genotype. Most phenotypes are affected by both the genotype and environment in which one has lived one's life including everything that has ever happened. We often consider "nature" as the unique genome which one carries and "nurture" as the environment in which one has lived one's life. The phenotype is dependent on the genetic makeup of the organism and influenced by the environment to which the organism is subjected across various epigenetic processes. It includes all of the organism's characteristics, including traits at multiple levels of biological organization, ranging from individual behavior and trait evolution through morphology, physiology, cellular characteristics, biochemical pathways, and gene expression.

A phenotypic trait (simply trait or character state) is a distinct variant of an organism's characteristic, and it is an obvious and measurable trait that is expressed in an observable way. For example, the eye color (green, blue, brown, and hazel) is a phenotypic trait which is a polygenetic inheritance. The phenotypic trait may be either inherited parentally or determined environmentally; that is to say, some traits are determined by the genotype, and some traits are determined by environmental factors. The different genes or alleles caused by mutation can be passed on to successive generations, resulting in different phenotypic traits. Though the environment can affect the phenotype, the heritability of a phenotypic trait is defined as the proportion of the total phenotypic variation of this specific trait that is elucidated only by the genetic variation [3]. The phenotypic variation of a trait (P) can be divided into three contributions as follows: latent genetic (G) factors, environmental (E) factors, and gene–environment interactions(G × E) [3].

If an organism inherits two same alleles, it is homozygous and expresses only one phenotypic trait. If an organism inherits two different alleles, it is heterozygous and may express more than one phenotypic trait. The phenotypic traits can be dominant or recessive. In completely dominant inheritance, only dominant traits are observable because the phenotype of the dominant trait entirely masks the phenotype of the recessive trait. In contrast, the dominant allele does not mask the other allele completely in incomplete dominance inheritance, resulting in a phenotype that is a mixture of traits from both alleles. In codominance relationships, both alleles are fully expressed, resulting in a phenotype in which both traits are demonstrated independently.

5. Conclusion

The gene expression function in eukaryotes, prokaryotes, and viruses is to generate macromolecules for constituting cellular components and exhibiting living functions. Its processes can be regulated in several steps including transcription, posttranscriptional modification, translation, and posttranslational modification. Regulation of gene expression is to modulate the levels of production and the timing of the functional product production. Control of gene expression is critical to allow cells to manufacture their functional products when cells need them; in turn, this gives cells the flexibility to adapt to a variable environment and respond to external signals, stimuli, and damages. Cellular structures and functions can also be controlled by gene expression regulation. Consequently, the regulation is crucial for cells to proliferate, differentiate, transport, metabolize, and repair. It is advantageous to the versatility, development, and adaptability of living organisms. A phenotypic trait, the expression of genes in an observable way, is an obvious and measurable trait. The phenotype is variable depending on the genetic makeup of the organism and also influenced by the surroundings to which the organism is subjected across its morphogenesis, including various epigenetic processes. We have to pay attention to the significance, mechanism, function, and characteristic of gene expression and phenotypic traits. Recently new findings and their implications could be discussed in the broadest context possible. Future studies, analyses, and applications of the gene expression and phenotypic traits should also be highlighted.

Author details

Yuan-Chuan Chen
Program in Comparative Biochemistry, University of California, Berkeley, CA, USA

*Address all correspondence to: yuchuan1022@gmail.com

References

[1] Natale F, Vivo M, Falco G, Angrisano T. Deciphering DNA methylation signatures of pancreatic cancer and pancreatitis. Clinical Epigenetics. 2019;**1**(1):132. DOI: 10.1186/s13148-019-0728-8

[2] Saenen ND, Martens DS, Neven KY, Alfano R, Bové H, Janssen BG, et al. Air pollution-induced placental alterations: An interplay of oxidative stress, epigenetics, and the aging phenotype? Clinical Epigenetics. 2019;**11**(1):124. DOI: 10.1186/s13148-019-0688

[3] Tiphaine CM, Jordana TB, Timothy DS. Twin studies and epigenetics. In: International Encyclopedia of the Social & Behavioral Sciences. 2nd ed. Elsevier, London, UK. 2015. pp. 683-702. DOI: 10.1016/B978-0-08-097086-8.82051-6

Section 2

Sex Determination

Chapter 2

Instability of Sex-Determining Systems in Frogs

Michihiko Ito

Abstract

All of the anuran amphibians examined so far have genetic sex-determining systems, which include female heterogametic ZZ/ZW and male heterogametic XX/XY types. For example, the Japanese wrinkled frog *Glandirana rugosa* has both types. Most of frog species including the African clawed frog *Xenopus laevis* possess homomorphic sex chromosomes, while most mammalian and avian species have heteromorphic sex chromosomes. Thus, there should be a variety of sex-determining genes and sex chromosomes in frogs, although only *X. laevis* W-linked gene *dm-W* has been reported as a sex-determining gene. Interestingly, estrogen or androgen can induce sex reversal in many frog species, suggesting a vital role of sex steroid hormones on sex identity. In other words, frogs in the same order are good examples for the understanding of diversity of sex-determining systems. In this chapter, I summarize the diversity of frog sex-determining systems and discuss why sex-determining genes and systems have been unstable in frogs.

Keywords: sex determination, sex chromosome, sex-determining gene, sex steroid, default sex, ectothermy

1. Introduction

Sexual reproduction is the most common life cycle in animals and plants. Meiotic recombination mediated through sexual reproduction is believed to allow genetic variation for survival of some populations against environmental changes. Thus, sex systems are very important for life evolution and biodiversity. In vertebrates, female and male sexes could be mainly defined by the property of gonads, ovaries producing eggs and testes producing sperm, respectively. Importantly, undifferentiated gonads in most vertebrate species have potential to differentiate into ovaries and testes. Then sex determination could be defined as the decision of bipotential gonads to develop as either ovaries or testes in vertebrates.

There are a variety of sex-determining systems in organisms. In vertebrates, they could be classified roughly into two types: genetic and environmental types. Endothermic vertebrates exclusively have the former system, which includes female (ZW) and male (XY) heterogametic sex chromosomes. Most mammalian and avian species have the XX/XY and ZZ/ZW systems, respectively, while there are both ZZ/ZW- and XX/XY-type systems in teleost fish, amphibians, and reptiles [1]. In addition, ectothermic vertebrates including reptiles and fish have not only the genetic sex-determining systems but also environmental sex-determining systems, such as temperature- and social-dependent types. Remarkably, all amphibian species possess the genetic systems, although they have ectothermic traits like reptiles and fish [1].

In the chapter, I introduce sex-determining systems, sex chromosomes, and sex-determining genes in amphibian frogs and discuss the relationships among them.

2. Sex-determining systems and sex chromosomes in frogs

As described in the above section, all anuran amphibians examined so far have the genetic sex-determining systems including the ZZ/ZW and XX/XY types (**Table 1**). For examples, the (African bullfrog) *Pyxicephalus adspersus*, African clawed frog *Xenopus laevis*, and the cane toad *Bufo marinus* have the ZZ/ZW type [2–5], while the African reed frog *Hyperolius viridiflavus* and the marsupial frog *Gastrotheca riobambae* adopt the XX/XY-type systems [6, 7]. Remarkably, the Japanese frog *Glandirana (Rana) rugosa* have five populations in Japan; their sex-determining systems include two ZZ/ZW and three XX/XY types [8].

Species	Sex-determining type	Morphology of sex chromosomes	Sex-determining gene
Xenopus laevis (African clawed frog)	ZZ/ZW	Homomorphic	W-specific *dm-W*
Pyxicephalus adspersus African bullfrog)	ZZ/ZW	Heteromorphic	
Glandirana rugosa (Japanese wrinkled frog)	ZZ/ZW or XX/XY	Heteromorphic/homomorphic	
Gastrotheca riobambae	XX/XY	Heteromorphic	
Hyperolius viridiflavus (African reed frog)	XX/XY	Homomorphic	

Table 1.
Sex-determining systems, sex chromosomes, and sex-determining genes in frogs.

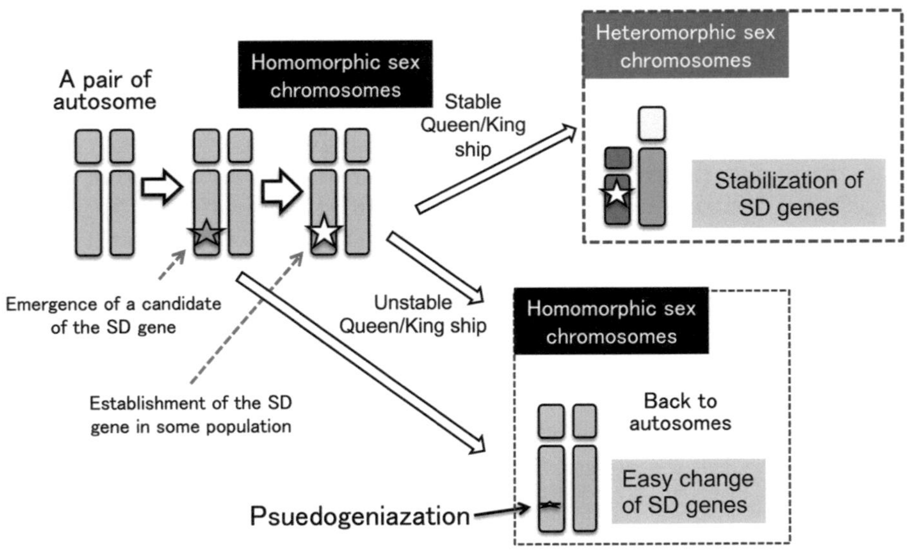

Figure 1.
A model for emergence and evolution of sex-determining genes and homomorphic and heteromorphic sex chromosomes in vertebrates. The model includes a proposal of "GENE-eat-GENE" model for changes of sex-determining genes in homomorphic sex chromosomes.

The XX/XY and ZZ/ZW systems in most mammals and all birds examined have been maintained for more than a 100 million years, which is greatly connected with the monophyletic and heteromorphic sex chromosomes among most species of therian mammals or avians: the monophyly of the Z or Y sex chromosomes is closely related to the maintainability of the sex-determining gene *Dmrt1* on the Z chromosome or *Sry* on the Y chromosome, respectively [1]. In contrast, more than 90% species of frogs including *X. laevis* have homomorphic sex chromosomes [9–11]. In fact, sex chromosome homomorphism is well conserved among many vertebrate species except for mammals and birds. In 2012, we proposed a hypothesis for the coevolution of sex chromosomes and sex-determining genes, in which homomorphic sex chromosomes easily allow changes of sex-determining genes, resulting in changes of sex chromosomes. On the contrary, highly differentiated heteromorphic sex chromosomes including mammalian XY and avian ZW chromosomes are easily maintained, resulting in a stable fixation of a particular sex-determining gene, because each sex chromosome has gained important functions except for sex determination ([12, 13]; **Figure 1**). This context could lead to the conclusion that there are a variety of sex-determining genes in frogs [1], although few amphibian sex-determining genes except for *dm-W* we discovered in *X. laevis* [14] have been identified yet.

3. Discovery of a female sex-determining gene *dm-W* in the African clawed frog

In 1990, human *SRY* was discovered as a sex-determining gene, which was the first report among vertebrate species [15], followed by mouse *Sry* [16]. Now *Sry* is believed to be a sex-determining gene in many species of therian mammals. After about 10 years, the second vertebrate sex-determining gene named *dmy* (also known as *dmrt1bY*) was reported in the teleost fish medaka *Oryzias latipes* [17, 18]. Both the two genes function as Y-linked male-determining genes in the XX/XY-type sex-determining systems. In 2008, we discovered a W-linked sex (female)-determining gene *dm-W* from the frog *X. laevis* having a ZZ/ZW type [5]; *dm-W* was the first report as the sex-determining gene among amphibian species or ZZ/ZW-type vertebrate species. Among sex-determining genes reported so far, the *dm-W* gene is unique in that the gene is female genome-specific (W-linked) and causes ovary formation [5, 19]. Both the *dmy* and *dm-W* genes emerged from the duplication of *dmrt1* independently during species diversity in genus *Oryzias* and *Xenopus*, respectively [12]. Next, Smith et al. (2010) reported that the Z-linked *dmrt1* gene is necessary for male sex determination in the chicken (*Gallus domesticus*) [20]. Here I should describe what protein is doublesex Mab-3-related transcription factor 1 (DMRT1). The protein including a DNA-binding domain, called "DM domain," functions in gonadal somatic cell masculinization and germ cell development in most vertebrates as transcription factors [21].

X. laevis is an allotetraploid species, whose ancestor might emerge by hybridization between two closely related *Xenopus* diploid species [22]. Therefore, there are two homoeologous L and S subgenome-derived genes in most of the genes in *X. laevis*. Partial duplication of S subgenome-derived *dmrt1* (*dmrt1.S*) leads to the emergence of *dm-W* [5, 23, 24]. In addition, we recently reported that *dm-W* evolved after allotetraploidization [24].

The DM domain of DM-W has about 90% amino acid sequence identity with those of DMRT1.L and DMRT1.S. However, the DM-W C-terminal region shares almost no similarity with those of DMRT1s. The last fourth exon of *dm-W* coding the C-terminal region emerged as a new exon [5]. We reported that DM-W and DMRT1 could cause primary ovarian and testicular formation in developing ZW and ZZ gonads, respectively [19], and proposed a sex-determining model for the ZZ/ZW type that DM-W determines female sex by antagonizing DMRT1; *dm-W* evolved from a masculinizing gene *dmrt1* as a dominant negative-type gene [14].

4. Sex reversal and sex chromosome differentiation

Although all frog species might genetically determine sex as mentioned above, most frog species could accept male-to-female or female-to-male sex reversals by treatment of sex steroids, estrogen, or androgen, respectively, during tadpole development [1]. Importantly, many frogs of them have homomorphic sex chromosomes. For example, *X. laevis* carries homomorphic W and Z sex chromosomes [5], and the estradiol-treated ZZ tadpoles developed to female adults [1]. In addition, we reported ZW female-to-male sex reversals in *X. laevis* transgenic tadpoles with *dm-W* knockdown or germline stem cell-specific knockdown of *dmrt1* and ZZ male-to-female sex reversals in *X. laevis* transgenic tadpoles carrying the *dm-W* expression plasmid [5, 19, 21].

Moreover, we recently analyzed detail structures of the sex chromosomes on 2Lq32-33 in *X. laevis*, revealing 278 kb W-specific region including three W-specific genes, the sex-determining gene *dm-W*, *scanw*, and *ccdc69w*, and 83 kb Z-specific region including one Z-specific gene *capn5z* [24]. Importantly, both gynogenetic WW and estrogen-driven sex-reversed ZZ individuals could develop into normal fertile females [25, 26]. These findings suggest that the homomorphic W/Z sex chromosomes in *X. laevis* are now differentiating but not so differentiated yet. In other words, *X. laevis* sex chromosomes have the potential to accept sex reversal and a new sex-determining gene.

5. Conclusions and perspective

All frogs examined possess genetic sex-determining systems, and most of them have homomorphic sex chromosomes. The genetic systems could be easy to change during species diversity, that is, the instability of the systems, maybe because of homomorphic sex chromosomes, which could have a potential to convert a sex-determining gene into a new one on another chromosome, resulting in the change of sex chromosomes. Then I propose a "GENE-eat-GENE" model for turnover of sex-determining genes: there has been battles among the present sex-determining gene and candidates of new sex-determining genes for king/queen ship in some populations holding homomorphic sex chromosomes (**Figure 1**). Accordingly, I predict that there are great many sex-determining genes in frogs, although only one *dm-W* has been identified as sex-determining genes. Frogs belong to the order Anura, which collects several thousands of species. Therefore they could be good examples for studying the relationships between sex-determining systems and species diversity.

Author details

Michihiko Ito
Department of Biosciences, School of Science, Kitasato University, Sagamihara, Japan

*Address all correspondence to: ito@sci.kitasato-u.ac.jp

References

[1] Ito M. Sex determination and differentiation in frogs. In: Kobayashi et al., editors. Reproductive and Developmental Strategies. Springer; 2018. pp. 349-366

[2] Abramyan J, Ezaz T, Graves JA, Koopman P. Z and W sex chromosomes in the cane toad (*Bufo marinus*). Chromosome Research. 2009;**17**:1015-1024

[3] Chang CY, Witschi E. Genic control and hormonal reversal of sex differentiation in *Xenopus*. Proceedings of the Society for Experimental Biology and Medicine. 1956;**93**:140-144

[4] Schmid M, Bachmann K. A frog with highly evolved sex chromosomes. Experientia. 1981;**37**:243-245

[5] Yoshimoto S, Okada E, Umemoto H, Tamura K, Uno Y, Nishida-Umehara C, et al. A W-linked DM-domain gene, DM-W, participates in primary ovary development in *Xenopus laevis*. Proceedings of the National Academy of Sciences of the United States of America. 2008;**105**:2469-2474

[6] De Almeida CG, Grafe TU, Guttenbach M, Schmid M. Karyotype and chromosome banding in the reed frog *Hyperolius viridiflavus ommatostictus* (Amphibia, *Anura*, *Hyperoliidae*). Experientia. 1990;**46**:509-511

[7] Schmid M, Haaf T, Geile B, Sims S. Chromosome banding in Amphibia. VIII. An unusual XY/XX-sex chromosome system in *Gastrotheca riobambae* (Anura, Hylidae). Chromosoma. 1983;**88**:69-82

[8] Miura I. An evolutionary witness: The frog rana rugosa underwent change of heterogametic sex from XY male to ZW female. Sexual Development. 2007;**1**:323-331

[9] Eggert C. Sex determination: The amphibian models. Reproduction, Nutrition, Development. 2004;**44**: 539-549

[10] Malcom JW, Kudra RS, Malone JH. The sex chromosomes of frogs: Variability and tolerance offer clues to genome evolution and function. Journal of Genomics. 2014;**2**:68-76

[11] Schmid M, Steinlein C, Bogart JP, Feichtinger W, León P, La Marca E, et al. The chromosomes of terraranan frogs. Insights into vertebrate cytogenetics. Cytogenetic and Genome Research. 2010;**130-131**:1-568

[12] Ito M, Mawaribuchi S. Molecular evolution of genes involved in vertebrate sex determination. In: eLS. Chichester: John Wiley & Sons, Ltd; 2013

[13] Mawaribuchi S, Yoshimoto S, Ohashi S, Takamatsu N, Ito M. Molecular evolution of vertebrate sex-determining genes. Chromosome Research. 2012;**20**:139-151

[14] Yoshimoto S, Ito M. A ZZ/ZW-type sex determination in *Xenopus laevis*. The FEBS Journal. 2011;**278**:1020-1026

[15] Sinclair AH, Berta P, Palmer MS, Hawkins JR, Griffiths BL, Smith MJ, et al. A gene from the human sex-determining region encodes a protein with homology to a conserved DNA-binding motif. Nature. 1990;**346**:240-244

[16] Koopman P, Gubbay J, Vivian N, Goodfellow P, Lovell-Badge R. Male development of chromosomally female mice transgenic for *Sry*. Nature. 1991;**351**:117-121

[17] Matsuda M, Nagahama Y, Shinomiya A, Sato T, Matsuda C, Kobayashi T, et al. DMY is a Y-specific DM-domain gene required for male development in the medaka fish. Nature. 2002;**417**:559-563

[18] Nanda I, Kondo M, Hornung U, Asakawa S, Winkler C, Shimizu A, et al. A duplicated copy of DMRT1 in the sex-determining region of the Y chromosome of the medaka, *Oryzias latipes*. Proceedings of the National Academy of Sciences of the United States of America. 2002;**99**:11778-11783

[19] Yoshimoto S, Ikeda N, Izutsu Y, Shiba T, Takamatsu N, Ito M. Opposite roles of DMRT1 and its W-linked paralogue, DM-W, in sexual dimorphism of *Xenopus laevis*: Implications of a ZZ/ZW-type sex-determining system. Development. 2010;**137**:2519-2526

[20] Smith CA, Roeszler KN, Ohnesorg T, Cummins DM, Farlie PG, Doran TJ, et al. The avian Z-linked gene DMRT1 is required for male sex determination in the chicken. Nature. 2009;**461**:267-271

[21] Mawaribuchi S, Musashijima M, Wada M, Izutsu Y, Kurakata E, Park MK, et al. Molecular evolution of two distinct dmrt1 promoters for germ and somatic cells in vertebrate gonads. Molecular Biology and Evolution. 2017;**34**:724-733

[22] Session AM, Uno Y, Kwon T, Chapman JA, Toyoda A, Takahashi S, et al. Genome evolution in the allotetraploid frog *Xenopus laevis*. Nature. 2016;**538**:336-343

[23] Bewick AJ, Anderson DW, Evans BJ. Evolution of the closely related, sex-related genes DM-W and DMRT1 in African clawed frogs (*Xenopus*). Evolution. 2011;**65**:698-712

[24] Mawaribuchi S, Takahashi S, Wada M, Uno Y, Matsuda Y, Kondo M, et al. Sex chromosome differentiation and the W- and Z-specific loci in *Xenopus laevis*. Developmental Biology. 2017;**426**:393-400

[25] Colombelli B, Thiebaud CH, Muller WP. Production of WW superfemales by diploid gynogenesis in *Xenopus laevis*. Molecular & General Genetics. 1984;**194**:57-59

[26] Villalpando I, Merchant-Larios H. Determination of the sensitive stages for gonadal sex-reversal in *Xenopus laevis* tadpoles. The International Journal of Developmental Biology. 1990;**34**:281-285

Chapter 3

Comparison of Sex Determination in Vertebrates (Nonmammals)

Aleksandr F. Smirnov and Antonina V. Trukhina

Abstract

The chapter is devoted to the consideration of sex determination in vertebrate groups of nonmammalians: fish, amphibians, reptiles, and birds. Attention is drawn to the fact that all these groups of animals, unlike mammals, are implemented hormonal control options for primary sex determination, and there is a possibility of sex reversion. Determination of gonadal development in vertebrates like testis or ovary was initially controlled mainly by sex hormones (fish and amphibians). Later, various sex determining genes were involved in this process. The system was quite plastic and was able to respond to changes in external conditions (reptiles). The appearance of heteromorphic sex chromosomes (birds) has led to the emergence of some specific W chromosomal signal, which provides estrogen control of the development of a heterogametic sex. In mammals, the control of the primary determination of sex (the appearance of the gonad) becomes purely genetic, and the role of sex hormones is reduced to the differentiation of testis or ovaries.

Keywords: sex determination, sex hormones, sex chromosomes, sex determining genes

1. Introduction

Gender is a set of morphological and physiological characteristics of the organism, providing reproduction, the essence of which is to fertilization, i.e. the fusion of male and female germ cells (gametes) in zygote, which develops into a new organism. Differentiation of sex (its phenotypic manifestation) includes two successive stages: the primary determination of sex and the appearance of secondary (external) sexual characteristics (actual differentiation). It is believed that the concept of this process is conservative. Sex determination is both a genetic and ecological process, with the sex of the individual being determined by an alternative physiological solution. It is assumed that there are two main mechanisms for determining sex: genetic (GSD—genetic sex determination) and environmental (ESD—environmental sex determination). Genetic sex is determined at the time of conception and depends on genetic differences between males and females, and ecological sex depends on external conditions in the absence of significant genetic differences and is determined after fertilization in response to environmental conditions. For birds and mammals, only the GSD is characteristic, and for crocodiles—TSD (one of the forms of ESD). In addition, there are two varieties of the genetic sex determination system: with heterogametic males (XY, mammals) and heterogametic females (ZW, birds). It should be noted that amphibians have both genetic systems, and for lizards, snakes, turtles, and bony fish, all possible variants of sex determination are described [1–3].

Sex steroid hormones including androgens, estrogens, and progesterone are present in all vertebrates which play essential roles in modulating a variety of behavior and processes, such as embryonic development, sexual differentiation, growth, aggression, reproduction, learning, memory, social communication, and so on. Many signaling actions of these sex steroid hormones are mediated by their receptors that belong to the superfamily of steroid nuclear receptors. Once a sex steroid hormone ligand binds to its receptor, the receptor becomes phosphorylated and is translocated into the nucleus, where it binds to specific DNA sequences and activates gene transcription. Androgens have a critical physiological role in reproductive biology and sexual differentiation, particularly in the development of male secondary sex characteristics [4, 5].

It is assumed that sex determination is a combination of hormonal and genetic factors and is divided conditionally into appropriate stages. This phenomenon is reflected in the possibility of sex inversion—the possibility of its complete or partial hormonal alteration. For fishes and amphibians, there is the sensitivity of normal development of the gonads to androgens and estrogens. In reptiles, birds and marsupials, only estrogens are effective. The appearance of the gonads of placental mammals does not depend on sex hormones. This trend is associated with the stability of growing offspring or incubation of eggs [6].

The proposed chapter will consider the system of sex determination in fish, amphibians, reptiles, and birds in comparing the role of hormonal and genetic mechanisms, possibilities, and mechanisms of sex inversion.

2. Features of sex determination in fishes

Fishes are perhaps the most complex group of animals in the mechanism of sex determination. Only bony fish include over 30,000 species. It is the largest group of vertebrates. They are divided into three groups in accordance with the laws of sex determination: (1) gonochoristic species whose sex is determined genetically or through environmental factors; (2) sequential hermaphrodites (about 2% of all existing species), changing the sex of males to females (protandrous), the sex of females to males (protogynous), or in both directions (serial) in the process of ontogenesis; (3) unisexual type of sex determination (characteristic only for Amazon mollies (*Poecilia formosa*)). Gonochoristic genetics of sex in fish is largely unclear. Functional hermaphroditism occurs in many different species of animals such as echinoderms, crustaceans, molluscs, and fish; however, it is lost in vertebrates during the transition from amphibians to mammals. From here, fishes provide a unique model for studying the mechanism of hermaphroditism in vertebrates. Unfortunately, only one species of fish (Japanese medaka—*Oryzias latipes*) was identified by a primary system of sex determination [7, 8].

The Japanese medaka (Oryzias latipes) and Maebashi medaka (*Oryzias curvinotus*)—species with heterogametic male sex with homomorphic sex chromosomes that are a very early stage of evolution, the recently described Y-chromosome plot, containing hypothetical gene *dmy*. This gene is specifically expressed in the gonads and is essential for embryo development in male type. Gene *dmrt1bY* (*dmy*) homologous (about 80%) of the *dmrt1* gene in other species of vertebrates represents the equivalent of *sry* gene in mammals. It is important that medaka *dmy*/*dmrt1bY* is a unique system. This species is described as ontology mammalian *sox9* gene, but in contrast to amniotes and amphibians, this does not play a role in determining the testes. Sex determination system of medaka is unstable. Medaka has interesting significant genetic divergence: *dmy* gene is absent in some lines of the Japanese medaka (over 10%) and other types of fish of the genus *Oryzias.* In some laboratory lines, the proportion of homogametic males (XX) exceeds 20%. It is believed that gene *dmy* has occurred as a result of the *dmrt1* gene duplication and transposition of part of its copy size to 280 kbp about 10 million years ago. The products of these genes differ only in one amino acid replacement (Ser26/Thr), which may have led to such differences by gender. It has been shown that the rate of synonymous substitutions in the *dmy* is 1.78 times greater than that of *dmrt1* and this is consistent with the hypothesis of evolution through males (male-driven evolution hypothesis). In birds and salmon, it has the same orientation. The speed ratio of nonsynonymous substitutions (dN) to synonymous (dS) is also higher in comparison with *dmy*/*dmrt1*.

Only two sex determining genes in vertebrates were described: *sry* and *dmy*. It is believed that the protein DMY performs two different functions in germ and somatic cells. In somatic cells surrounding germ ones, it affects the proliferation of the latter (for example, influencing a cascade of genes involved in the transmission of the estrogen signal). Another feature is the induction of development of pre-Sertoli cells (cells surrounding the primary germ cells (PGCs)) in the gonad heterogametic XY sex. In this case, there is an analogy with *sry*, which is involved in the activation of other genes that support the development of Sertoli cells. In medaka, there are other female-specific genes and male-specific genes (**Figure 1**). Moreover, the latter gene is located in autosomes. Some ideas of the diversity of sex determining genes among medaka given.

In this species, the sex determined region of the Y chromosome is only 260 kb (1% of the total length of the Y chromosome (59 Mbps)). In this area, there is suppression of recombination. In medaka, all XY individuals carry mutations in the gene *dmy* form ovaries. In individuals with altered *gsdf*-gene, sex inversion is also observed. It is believed that for medaka, the normal gene *dmrt1* (*dmy*) initiates the formation of the testes and controls their maintenance with *gsdf*. The study of sex chromosomes in six species of medaka from the group *celebensis* with XX/XY-sex determination showed that *O. marmoratus* and *O. profundicola* sex chromosomes homologous sex chromosomes *of O. latipes* from the LG10 linkage group. Four species *O. celebensis, O. matanensis, O. wolasi* and *O. woworae* marked homology

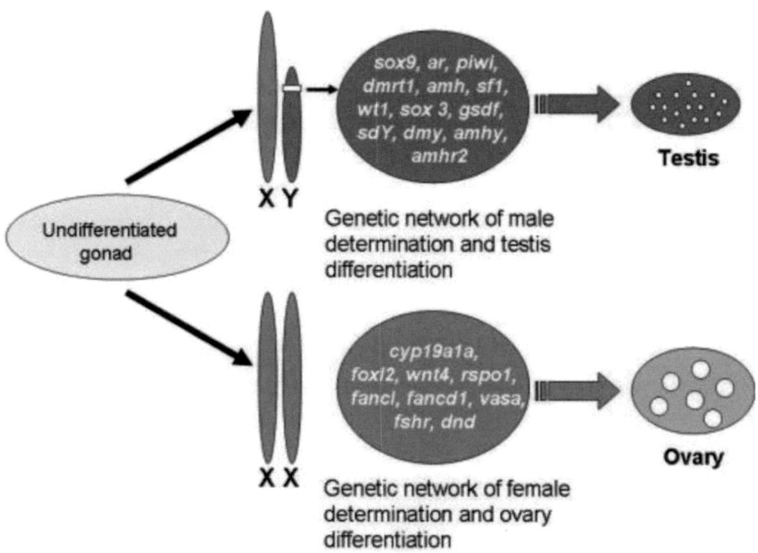

Figure 1.
A schematic diagram of sex determination and gonad (testis or ovary) differentiation in fish with XX/XY sex determination system (adopted from Mei and Gui [10]).

with the chromosomes of LG 24, which involves the transformation of chromosomes from *O. latipes* LG to 24 LG10 within this group. All six studied species share a common sex determined gene (SD). It is shown that genomic predecessor is the Y-chromosomal gene *sox3* and this process involves specific insertion (430 bp).

The zebrafish testes derived from *dmrt1* mutant fish fail to express the anti-Müllerian hormone (*amh*) gene, a key testis-expressed gene, and over-express the ovary-associated gene *foxl2*. Therefore, zebrafish *dmrt1* shares similar roles in male sexual development as other organisms in regulating sex determination and testis differentiation.

In other fishes, e.g., salmonids, there appear to be an early stage of differentiation of sex chromosomes. In rainbow trout (*Oncorhynchus mykiss*) with monofactorial XX/XY system of sex determination, a new gene *sdY* responsible for the development of testes is described. This gene is partially similar to the gene regulator of interferon 9. It has been found that highly conserved in *sdY* salmon is male Y-chromosomal gene for the majority of these species. It is assumed that it is the main testis determining gene for this group of fishes. For the two species of whitefish (subfamily *Coregoninae*), the *sdY* gene is found in both males and females. This implies that there is an alternative system of sex determination in this family. Among other candidate genes for sex determination, gene antimullerian hormone (*amh*) tilapia is discussed. Fishes with hermaphrodite sex determination (*Labridae*, fish-clowns—amphiprion (*Amphiprion),* and gobies—*Trimma okinawae*) have got bisexual gonads capable of restructuring with the participation of aromatase and gonadotropin receptors. For some species, such as blue tilapia (*Oreochromis aureus*), sex determined putative gene is located on the genetic map of a sex determining region consisting of more than 550 minisatellite markers [7, 9].

In vertebrates, until recently, only four sex determining genes were discovered: *sry* (in mammals), *dmrt1* (in domestic chicken), *dmy* (the Japanese medaka), and *dm-w* (the frog). Recently, four candidate genes were found for this role (and all fish): Patagonian aterin have *amhy*, Luzon ricefish (*Oryzias luzonensis*) have *gsdf*, and puffer (*Takifugu or Fugu*)—*amhr2* and rainbow trout—*sdy*. In the Nile tilapia (*Oreochromis niloticus)* gene *gdf*, (gonadal soma derived factor (*gsdf*)) also induces the development of the testes. Assume that the Atlantic salmon *sdY* gene product activates genes *gsdf* and *amh/mi*, thereby reducing the activity of aromatase (*cyp19a* gene), leading to the appearance of males. Where *sdY* is missing, aromatase is synthesized in quantities sufficient for the emergence of the females [8, 10].

Sex determining genes in fish are not conservative. It is believed that the reason for this is the more frequent variation of sex chromosomes in fish than other cold-blooded animals and mammals (**Figure 1**).

These objects sex determination has a high plasticity and is, therefore, possible sex reversal, even in species with established regulatory genes. Striped Danio (*Danio rerio*) experimental data are in good agreement with polygenic sex determination (PSD) when the sex is determined by allelic combinations of several loci. Typically, these loci are dispersed throughout the genome, but some species of bony fish are placed in special sex chromosomes. In hermaphroditic fish, ovotestis develops first, and then secondary sex determination occurs. So, the black bass individuals (genus *Micropterus*) in the first 2 years of life are males, but in the third year, 50% of them are transformed into females. Sex determining male genes such as *dmrt1*, *amh*, and *amhr2* are activated during differentiation of the testis, and their expression is maintained at high level during the period of functioning as males. High dose estrogen E2 induces the development of ovarian and testicular tissue degradation [11, 12].

In fish, there are two systems of sex determination: XX/XY and ZW/ZZ. The most common one is the last. Exploring the flatfish *Cynoglossus semilaevis* as a model species with genetic sex determination system of ZW-type and the simultaneous presence of ESD, it was found that about 14% of females at a temperature of 22°C become males (pseudomales). It is believed that there is *dmrt1* gene (double sex and mab-3 related transcription factor 1) which is the sex determining gene in this species. It was also shown that pseudomales change the level of methylation of a certain portion of the Z chromosome, resulting in the intensity of transcription in this area as in normal males. In females, on the contrary, the activity of the corresponding plot of W chromosome by methylation is suppressed. Unusual WXZ-system is described for the swordtail (*Xiphophorus helleri*). Not so many fish species had morphologically different sex chromosomes (about 10%) and in most species they are in the early stages of their differentiation. For many members of this class, sex is determined by the environment, and even changes under the influence of behavioral factors. There are species with heterogametic male and female [13].

Fish is characterized by plasticity of germ and somatic cells. This plasticity is maintained throughout the life cycle. Furthermore, they have described the influence of factors on this process such as temperature, pH, density of population, etc. It should be noted that the temperature sensitivity of fish is different from that of reptiles, especially because these types of monosexual populations are rare, even under extreme conditions. TSD in fish is less common than previously thought. The effect of estrogens, acting via estrogen receptors (ER) and directly or indirectly regulating P450arom and AMH, is particularly noticeable. It is noted that the analysis of the differences between gonochoristic and hermaphroditic fish species will help to understand the mechanism of plasticity of sex determination in vertebrates. In addition, there is the idea that gender in fish depending

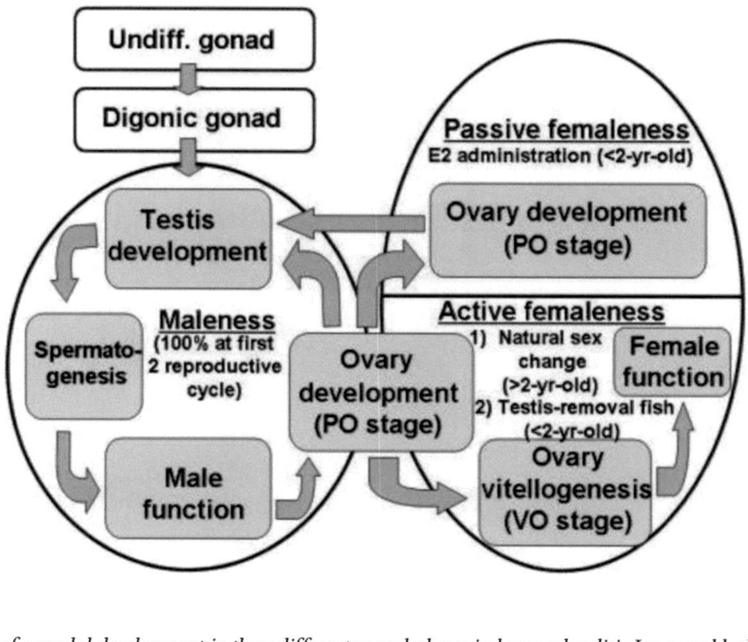

Figure 2.
The profiles of gonadal development in three different sexual phases in hermaphroditic Japanese black porgy, Acanthopagrus schlegelii. Maleness: the fishes are functional males in the first two spawning seasons. The testis exists at all stages of the reproductive cycles in maleness. Active femaleness: the fishes are functional females following the natural sex change that occurs in fish older than 2 years or when induced by the removal of the testis of the digonic gonad. The ovary could reach to the stage of vitellogenesis, vitellogenic, and mature oocytes. Passive femaleness: Long-term E2 (4–6 mg/kg feed) administration for 2–3 months results in the appearance of a dominant ovary (with the primary oocytes) with a regressed testis in fish younger than 2 years old, and no vitellogenic oocytes are observed in E2-induced sex-changing fish. A reversible sex change (from passive femaleness to maleness) exists after E2 administration has been withdrawn. Undiff. gonad, undifferentiated gonad; E2, estradiol-17β; PO, primary oocyte stage; and VO, vitellogenic oocyte stage (adopted from Wu and Chang [16]).

on species is a complex trait under the control of one or many genetic factors in addition to environmental effects [9, 14]. In the Chinese tongue sole (*Cynoglossus semilaevis*), genetic ZZ females may change into pseudomales, thereby increasing aquaculture costs because of the lower growth rate of the males than that of the females. A new locus was identified to regulate sex reversal interactively with the SNPCyn_Z_6676874; the linkage between these two loci and the absence of W sperm for pseudomales clearly elucidate the genetic architecture of sex reversal in the tongue sole [15]. Sexual determination in zebrafish is unique in that laboratory strains lack a sex chromosome, and no sex determining gene has been identified. GPER (estrogen receptor) is not required for normal sex differentiation, gonad development, or gonad function in zebrafish [16]. Genetic studies suggest that gonadal sexual fate is not only established by competition for primacy between two sexes via antagonistic signaling pathways during embryonic development but also requires active maintenance to suppress the opposite sex during adulthood. Documented in about 2% of teleost species spanning over 20 families, functional sex change generally occurs in three ways: protogynous (female-to-male), protandrous (male-to-female), and sequentially bi-directional. Most sequentially hermaphroditic fish are protogynous. Sex change in all hermaphroditic species involves radical gonadal transformation, and follows diverse ontogenetic pathways in different lineages particularly where sequential hermaphroditism has independently evolved. Gonadal transition in sex-changing fish is accompanied by changes in plasma concentrations of gonadal steroids. These steroids control gonad differentiation and maintain sexual phenotypes in teleost fish, wherein 17β-estradiol (E2) and 11-ketotestosterone (11-KT) function as the major estrogen and androgen, respectively. The balance between estrogen and androgen production is expected to control sexual fate of the gonads during sex change. For example,

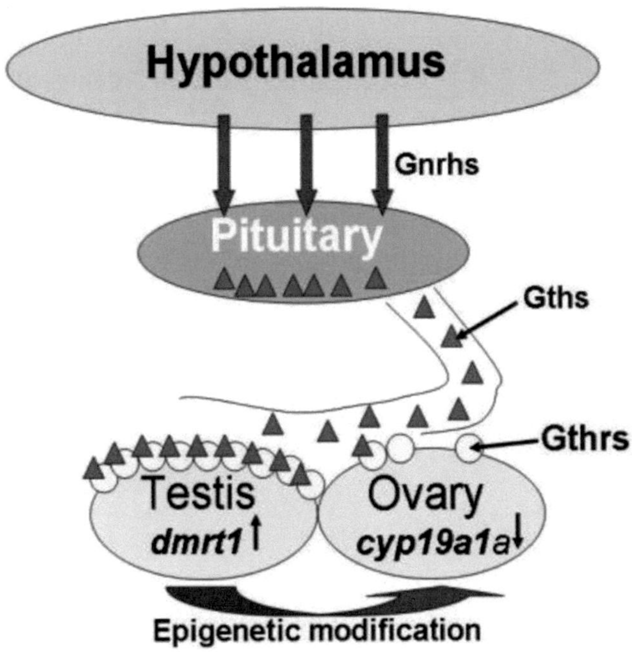

Figure 3.
The potential mechanism for sexual fate decision through the Gnrhs—Gths—Dmrt1 axis (brain-pituitary-testis axis). The model shows that the male fate decision is controlled by gonadotropins through the Gnrhs—Gths—Dmrt1 axis. The testis may stimulate the epigenetic modification of the ovary by DNA methylation of the cyp19a1a promoter to suppress the cyp19a1a expression. Gnrhs, gonadotropin-releasing hormones; Gths, gonadotropins; and Gthrs, gonadotropin receptors (adopted from Wu and Chang [16]).

factors regulating *cyp19a1a* expression are strong candidates for the trigger that initiates gonadal sex change; *cyp19a1a* promoter regions contain binding motifs for numerous factors that potentially regulate its expression [15].

However, hermaphroditic fishes have a plastic sex, and a stable sex is difficult to maintain with sex steroids. The black porgy regulated the dynamic development of both sexes; only one sex can grow while the other sex exists in a rudimentary stage (**Figure 2**). The sexual fate of the digonic gonad is determined by the male fate maintenance and through the Gnrh—Gth—Dmrt1 signaling. Altogether, testicular *dmrt1* and ovarian *cyp19a1a* expression are critical to the sexual fate of a male phase and female phase, respectively (**Figure 3**).

3. Sex in amphibians

Amphibians have two sex determined systems: XX/XY and ZZ/ZW. Most tailed amphibians (order *Caudata*) have XX/XY-system. For 63 species of 1500, sex was determined and only 20 species have differing sex chromosomes. Males of some New Zealand frogs (*Leiopelma hamiltoni* and *L. hochstetteri*) have heterogametic sex. In most amphibians, sex chromosomes are homomorphic (undifferentiated) in both sexes and are characterized by frequent turnover. This is in sharp contrast to sex chromosomes in two major vertebrate groups, the mammals and birds, where they are heteromorphic in one sex and are highly conserved. Thus, amphibians are excellent research materials on the turnover of sex sensitive to a resistant state, indicating the relationship between sex chromosome turnover and sex ratio control.

Models of sex differentiation in amphibians can be divided into three types: (1) a direct development of the undifferentiated gonads into testes or ovaries, (2) the development of the undifferentiated gonad into the ovary and subsequent development of the testis through the ovary, and (3) the development of the testes through the intersex phase (prodifferentiating type) [17]. For a long time, genes that determine sex could not be found in amphibians. Recently, for smooth clawed frog (*Xenopus laevis*), the candidate gene has been found suitable for such a role. It is believed that it is involved in the development of the ovary. African clawed frog has a ZZ/ZW system of sex determination. Its *dm-w* gene was described. It is localized in the X chromosome and possessed a DM-domain. The nucleotide sequence of gene encoding a DNA-binding domain has 89% identity with *dmrt1*, but there is not similarity in transactivational region *dm-w* and *dmrt1*: genes are expressed exclusively in the primordial gonads, and *dm-w* is expressed more actively than in the gonads of ZW-larvae. The gene *dmrt1* (*dmrt1α* and *dmrt1β*) is located in autosome and there are no differences in its expression in males and females. The product of this gene enhances the expression of *cyp19* and *foxl2* ones. A similar gene was not detected in other species of amphibians. It is assumed that in these frogs, homo- and heterodimer products of *dmrt1* and *dm-w* participate in the sex determination [18, 19] (**Figure 4**).

In the northern crested newt (*Triturus cristatus*), the proportion of males increases when the ambient temperature increases, and a decrease of temperature leads to an excess of females. Thus, in amphibians, an increase or decrease of the ambient temperature leads to a modification of the normal development of the gonads and sex determination. Here, sex-determining genes are not the decisive factor in determining sex. A number of experiments have shown that atrazine and some other pesticides that affect the endocrine system affect the formation of sex in frogs. As a result, males are changed to females. Exogenous steroids (introduced from the outside) are also changing the sex in amphibians [20]. The unexplainable mechanism of sex determination in the rice frog species was introduced. Amphibians bearing a novel sex determining mechanism are yet to be identified [17].

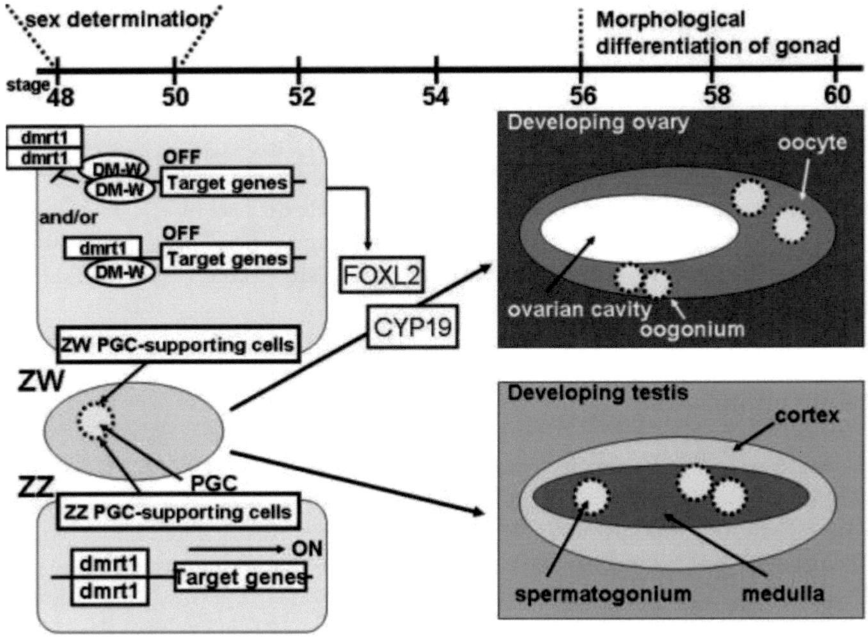

Figure 4.
Model of ZZ/ZW-sex determined system and the formation of the ovary from Xenopus laevis (adopted from Liu et al. [18]).

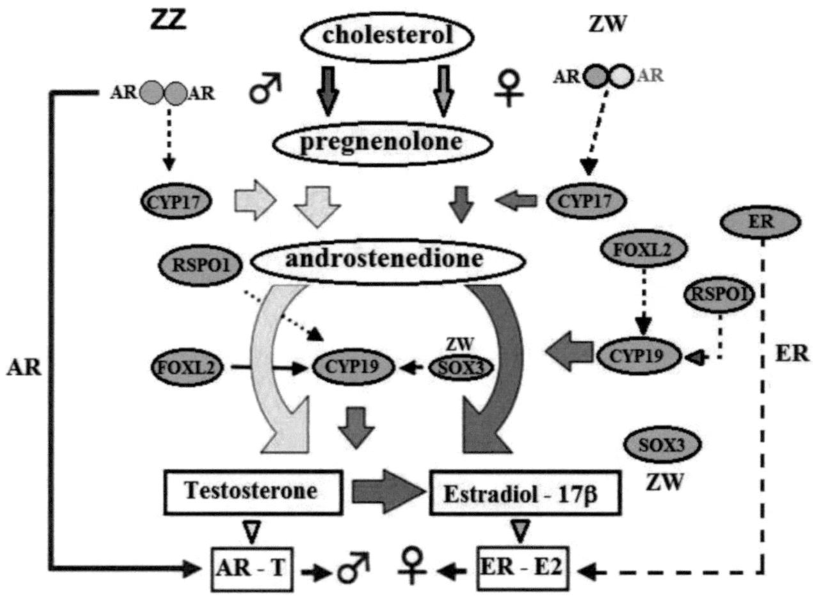

Figure 5.
The role of steroid hormones in sex determination from Rana rugosa. At the stage of sex determination in the undifferentiated gonads of males, testosterone is synthesized at the same time females synthesize estradiol-17β. Letters ZZ, ZW indicate sex chromosomes. AR-T and ER-E2 represent complex androgen receptor (AR) to testosterone, and respectively, estrogen receptor and estradiol-17β (adopted from Nakamura [23]).

The dominant hypothesis of sex determination for amphibians is proposed in relation to the *Rana rugosa.* In the Japanese wrinkled frog (Rana rugosa), four populations are described, in one of which (the northern population) females are heterogametic. Assume that sex determining genes really do not need to determine the sex of amphibians, as well as the presence of the transcription factor, localized in the X or W chromosomes, influencing the feminization of vertebrates with TSD or GSD systems of sex determination. In males, if there is a specific mechanism of sex determination, it is likely that it supports the regulation of steroid hormones in undifferentiated gonads through the inhibition of *cyp19* gene transcription for the formation of the ovaries. In the scheme of **Figure 5**, a possible role of steroid hormones in sex determination is shown for *Rana rugosa* [21]. According to the next experimental data, complete female-to-male sex reversal occurred in the AR-Tg-transgenic ZW female frogs when a low dosage of T was supplied in the rearing water of tadpoles. In the sex reversed testes, the expression of *dmrt1*, *ar*, and *cyp17* genes required for masculinization was significantly upregulated. Next, AR-knockdown (KD) ZW female frogs were produced by the CRISPR/Cas9 system. Interestingly, no sex-reversal was observed in AR-KD ZW female frogs when the gonads were treated with dosages of T high enough to induce complete female-to-male sex-reversal, even in wild type frogs. In the AR-KD ZW female gonads, the expression of genes required for masculinization was not up-regulated. These results indicate that AR together with androgens can be a male sex-determinant in an amphibian species [22, 23].

4. Sex in reptiles: determination of sex under the influence of temperature

Sex determination by environmental factors is mainly known in reptiles. The most well studied temperature sex determination (TSD) is occurring in three of the five main taxonomic groups of reptiles: turtles, crocodiles, and lizards, but it is not found in snakes. The adaptive significance of such sex determination mechanism is shown. During early embryonic development of gonad, epithelial cells are divided and unite in the epidermal strip of mesonephros mesenchyme. Further, during the so-called temperature-dependent period under the level of endogenous estrogen, such strip forms seminiferous tubules with Sertoli cell epithelium or gaps with squamous epithelium. The mechanism of this sex determination is poorly understood. Obviously, it is found in species with undifferentiated Y chromosome. The transition from the female promoting temperature (FPT) to male promoting temperature (MPT) is carried out in a temperature-period (TSP), during the so-called "window" of vulnerability [24].

In some species of reptiles, GSD is not fixed for life, and the original gender may change during development without changing the genotype. This phenomenon is known as environmental sex reversal (ESR) and observed also in insects, fish, and amphibians [25] (**Figure 6**).

In reptiles, there is an "open" sex determination program that is different from a "closed" program, characteristic of birds and mammals. It is believed that in this case, the gender depends on the ratio of estrogens and androgens during sexual differentiation of the gonads. The temperature of incubation may change the activity of genes encoding aromatase, estrogen receptor, and reductase. It is not excluded that different taxonomic groups of animals with TSD have different mechanisms of regulation of sex. There may be temperature-sensitive genes *sox9* and *dax1* (freshwater turtles—*Emydidae*) and genes *sox9*, *sf1*, and *wt1* (*Testudinidae*). In mammals, this mechanism is not valid, because the Y chromosome has genes that inhibit the aromatase enzyme.

For Mississippi alligator (*Alligator mississippiensis*), pond slider turtle (*Trachemys scripta*), and olive ridley (*Lepidochelys olivacea*, from the family of sea turtles), the expression level of the gene *dmrt1* was higher during the incubation of embryos at a temperature that contributed to the emergence of males. In reptiles and in particular

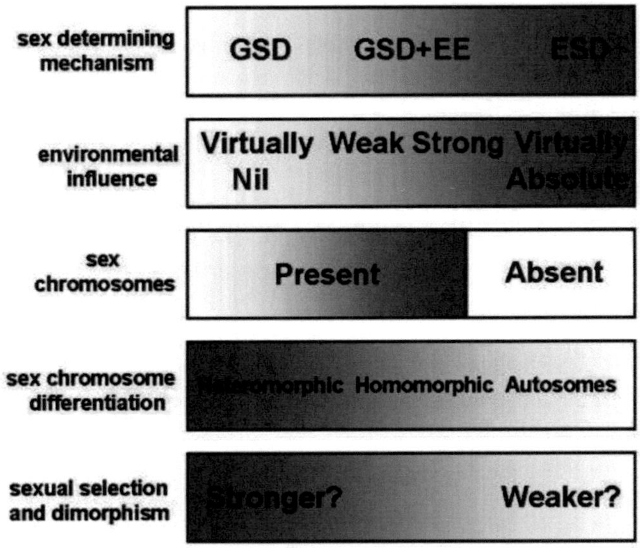

Figure 6.
The continuum of sex determination. Distribution mechanisms from GSD to ESD, including intermediate system to overcome genetic sex determination with environmental factors (GSD + EE) (adopted from Valenzuela et al. [26]).

Trachemys scripta elegans, a large amount of the KDM6B product is observed at a temperature favorable for males (MPT) and activates the expression of the *dmrt1* gene, and its reduction represses the expression of *dmrt1* and promotes the appearance of females. The latter is associated with H3K27 trimethylation. KDM6B is a member of the Jumonji gene family. It is believed that such genes are somehow regulated. One such regulator—*cirbp* (cold-inducible RNA binding protein)—has recently been described in the turtle *Chelydra serpentina* [27]. It managed to detect differences in the structure of *dmrt1*-gene in 34 species of reptiles with temperature and genetic mechanisms of sex determination, affecting sequence in exon 2 near DM-binding domain. In species with TSD, threonine occurs at position 54 (T54) and serine at position 57 (S57), while in species with a genetic sex determination mechanism, serine is observed in the S54-S57 position. This is obviously only the discovery of the molecular differences in sex determining gene when changing the mechanism of sex determination [28]. The discovery of the triploid male (ZZW) in the colubrid snake testifies to the absence of a particular role of the B chromosome in the determination of sex in this species [29].

Sex reversal has not yet been demonstrated in nature for any amniote, although it occurs in fish and rarely in amphibians. There is only one report about sex change in reptiles in the wild (Australian bearded dragon (*Pogona vitticeps*)) and the use of animals with inverse sex in order to experimentally induce a rapid transition from GSD to ESD. Controlled mating of normal males to sex-reversed females produces a viable and fertile offspring whose phenotypic sex is determined solely by temperature (temperature-dependent sex determination). The W sex chromosome is eliminated from this lineage in the first generation which indicates its specific role in genetic sex [30].

5. Sex determination in bird

In birds, estrogens play an important role in sex determination. They regulate expression of key sex determining genes during the first 3 days of embryonic development and further. At the same time, the set of sex chromosomes is equally important. Embryos with two Z chromosomes in birds develop as males, and those with ZW chromosomes develop as females. At present, two hypotheses on sex determination in birds compete. One of these hypotheses considers the number of Z chromosomes as a key sex determining factor, while the other hypothesis supposes the presence in W chromosome of the key gene controlling ovarian development or suppressing the appearance of testes. The presence in Z chromosome of a strong candidate gene for sex determination (DMRT1 gene) supports the dose scheme. **Figure** 7 presents a hypothetical scheme of genetic control of primary sex differentiation in *Gallus gallus.* The gonad appears on the 3.5th day (stage 22) as thickening on the surface of mesonephros. It consists of the epithelial layer of somatic and germ cells and medullary cordate layer (epithelial cords), which is mixed with mesenchymal cells. On the 6.5th day (30th stage), the first sex determining genes are activated. In the modern scheme of the genetic control of sex determination in birds (practically within the dose scheme), an epigenetic mechanism for switching off the single allele of avian key sex determining *dmrt1* gene in females through hypermethylation and using noncoding MHM RNA came into sharp focus (**Figure 8**) [30–34]. Synthetic aromatase inhibitors (an enzyme catalyzing the synthesis of estrogens) can induce steady female → male sex inversion. In this case, the left gonad becomes an ovotestis, or a testis, and the right gonad becomes a testis. Injection of aromatase inhibitors *in ovo* in most experiments was carried out on the third or fourth day of incubation. At the same time, in experimental males, injection of estradiol results in reversible feminization of the gonads [35, 36]. Unfortunately, the genetic and

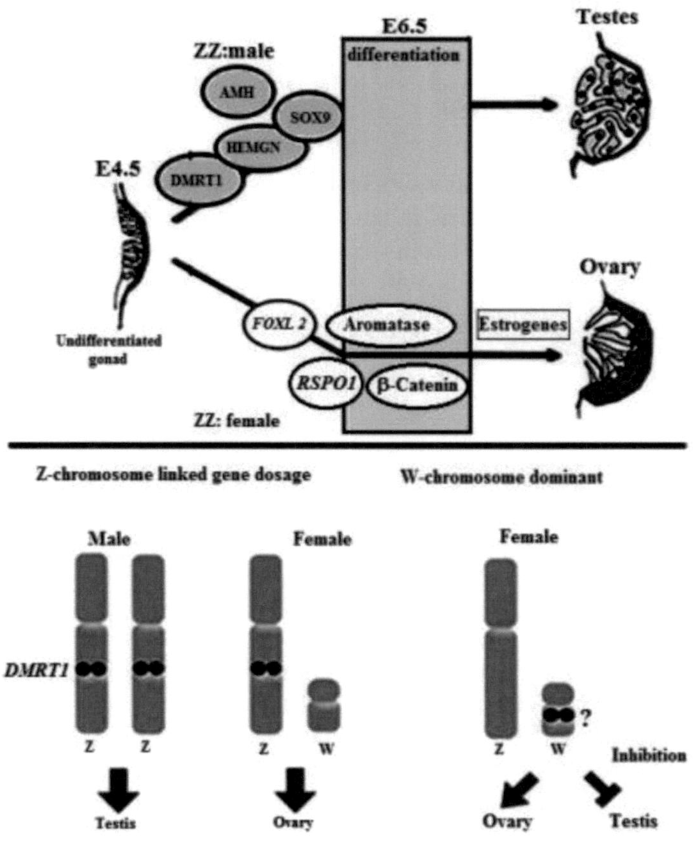

Figure 7.
Possible models of primary sex determination in birds by the example of Gallus gallus (adopted from Kuroiwa [33]).

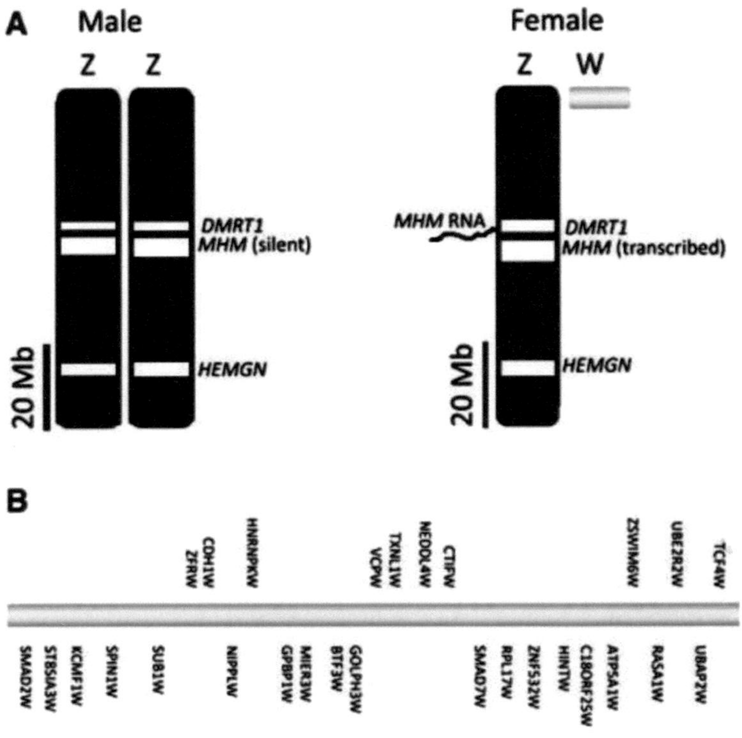

Figure 8.
The large Z chromosome (82.3 Mb) is drawn to scale next to the degenerate W chromosome (7 Mb). (A) Male (ZZ) have two copies of DMRT1 and HEMGN, while the female (ZW) only has one. The MHM locus is transcribed from the single Z in the female and may play a role in local dosage and epigenetic regulation of DMRT1 in the female. (B) Location and orientation of the 28 protein coding genes that are located on the W chromosome (adopted from Hirst et al. [34]).

hormonal status of individuals with sex inversion was not investigated. The two enzymes required for the synthesis of estrogen, aromatase, and 17-beta-hydroxysteroid dehydrogenase (17β-HSD) are synthesized only in ovarian medullary cords at the onset of morphological differentiation. It is suggested that the earliest expression of aromatase in birds is detected only on the fifth day of embryonic development. It is worth mentioning that the appearance of aromatase was recently demonstrated as early as in the maternal body, upon oogenesis in the theca layer of early follicles [37]. The data obtained make it possible to suggest earlier appearance of aromatase and estrogens in female gonadogenesis than that follows from the classical scheme of primary sex determination in *Gallus gallus* [38].

In birds, sex determination depends on sex hormones and sex-hormone-specific receptors. Estrogen receptors are also important in this process. In a recent study, the gonads and endocrine profile of a gynadromorphic chicken were described. It had male features on the right and female features on the left. At sexual maturity, the gonads of this bird were largely testicular. The right gonad was a testis, with $SOX9^{+}$Sertoli cells, $DMRT1^{+}$germ cells, and active spermatogenesis. According to histology, the left gonad was primarily testicular, but with a few number of peripheral aromatase follicles. The gynandromorph had low levels of serum 17β-estradiol (39 pmol/L). In contrast, the gynandromorph had very elevated levels of serum testosterone (41.3 nmol/L). Despite the elevated testosterone, the bird was female on one side of the body. The right male side was almost entirely ZZ (96%), whereas those from the left female side were a mixture of male (77% ZZ) and female (23% ZW) cells. It had a low percentage of ZW cells on the female side, but still had female sex-linked feathering, smaller muscle mass, smaller leg and spur, and smaller wattle. This indicates that sexually dimorphic structures such as the wattle, spur, and feathering must be at least partly independent of sex steroid effects. Even a small percentage of ZW cells appear sufficient to support female-type sexual differentiation [39–41]. Studies of chimeric embryos also support the hypothesis that avian sexual differentiation is largely, or partly, cell autonomous, involving direct genetic factors and steroid hormones.

6. Conclusion

So, estrogens and androgens play important roles in sexual differentiation and reproduction, particularly in the development and expression of male and female sexual characteristics. These effects are principally mediated by the estrogen and androgen receptors (ESRs and ARs), which belong to superfamily of the nuclear receptors [42]. The nature of the relationship between sex hormones and gender determining genes and the patterns of their interaction remains unclear. For some amphibians, the absence of appropriate genes and the replacement by control factors of steroid hormones and receptors are postulated. For birds, we can assume a special role of heteromorphism of sex chromosome and the presence of a specific interaction of the W and Z chromosomes. In this regard, we should mention the phenomenon of detection of specific chromosomes (germ line restricted chromosomes, GRS) found only in the germ cells of songbirds.

In mammals, aromatase is expressed later in embryonic development and the gonadal sex is formed independently of sex hormones and differentiation can occur in the absence of steroidogenesis. For mammals, two-step primary sex determination is typical. At the first stage, its determination is carried out by the *sry* gene. At the second one, sex hormones are synthesized in gonads and genetic endocrine regulation of sex development is maintained. It raises questions about the sensitivity to androgens and estrogens of sex determination in fish, amphibians, reptiles, and birds. The functional role of the emerging chromosome heteromorphism is not clear. It is believed that the realization of the phenomenon of sex reversal is different in nonmammal vertebrates and mammals. It is intended to introduce a special term for nonmammal's sex change [43].

So, determination of gonadal development in vertebrates like testis or ovary was initially controlled mainly by sex hormones (fish and amphibians). Later, various sex determining genes were involved in this process. The system was quite plastic and was able to respond to changes in external conditions (reptiles). The appearance of heteromorphic sex chromosomes (birds) has led to the emergence of some specific W chromosomal signal, which provides estrogen control of the development of a heterogametic sex. In mammals, the control of the primary determination of sex (the appearance of the gonad) becomes purely genetic, and the role of sex hormones is reduced to the differentiation of testis or ovaries.

Acknowledgements

This research was supported by a grant 17-04-01321A Russian Foundation for Basic Research (RFBR). The authors declare that the research was conducted in the absence of any commercial or financial relationships that could be considered as a potential conflict of interest.

Author details

Aleksandr F. Smirnov and Antonina V. Trukhina*
Department of Genetics and Biotechnology, Saint Petersburg State University, Saint Petersburg, Russia

*Address all correspondence to: a.trukhina@spbu.ru

References

[1] Johnson Pokorná M, Kratochvíl L. What was the ancestral sex-determining mechanism in amniote vertebrates? Biological Reviews of the Cambridge Philosophical Society. 2016;**91**(1):1-12. DOI: 10.1111/brv.12156

[2] Schärer L. The varied ways of being male and female. Molecular Reproduction and Development. 2017;**84**:94-104. DOI: 10.1002/mrd.22775

[3] Smirnov AF, Trukhina AV. Molecular-Genetic Mechanisms of Sex Determination in Animals. Scientific Research Publishing: An Academic Publisher; 2017. p. 118. ISBN: 978-1-61896-390-1

[4] Murashima A, Kishigami S, Thomson A, et al. Androgens and mammalian male reproductive tract development. Biochimica et Biophysica Acta. 2015;**1849**(2):163-170. DOI: 10.1016/j.bbagrm.2014.05.020

[5] Morohashi K, Baba T, Tanaka M. Steroid hormones and the development of reproductive organs. Sexual Development. 2013;**7**(1-3):61-79. DOI: 10.1159/000342272

[6] Tagirov MT. Sex determination and control mechanisms in birds. Biotechnologia Acta. 2013;**6**(1):62-72. DOI: 10.15407/biotech6.01.062

[7] Guerrero-Estévez S, Moreno-Mendoza N. Sexual determination and differentiation in teleost fish. Reviews in Fish Biology and Fisheries. 2009;**20**(1):101-121. DOI: 10.1007/s11160-009-9123-4

[8] Hattori RS, Strüssmann CA, Fernandino JI, et al. Genotypic sex determination in teleosts: Insights from the testis-determining *amhy* gene. General and Comparative Endocrinology. 2013;**192**:55-59. DOI: 10.1016/j.ygcen.2013.03.019

[9] Martínez P, Viñas AM, Sánchez L, et al. Genetic architecture of sex determination in fish: Applications to sex ratio control in aquaculture. Frontiers in Genetics. 2014;**5**:340. DOI: 10.3389/fgene.2014.00340

[10] Mei J, Gui JF. Genetic basis and biotechnological manipulation of sexual dimorphism and sex determination in fish. Science China. Life Sciences. 2015;**58**(2):124-136. DOI: 10.1007/s11427-014-4797-9

[11] Kikuchi K, Hamaguchi S. Novel sex-determining genes in fish and sex chromosome evolution. Developmental Dynamics. 2013;**242**(4):339-353. DOI: 10.1002/dvdy.23927

[12] Kobayashi Y, Nagahama Y, Nakamura M. Diversity and plasticity of sex determination and differentiation in fishes. Sexual Development. 2013;**7**(1-3):115-125. DOI: 10.1159/000342009

[13] Shao C, Li Q, Chen S, et al. Epigenetic modification and inheritance in sexual reversal of fish. Genome Research. 2014;**24**(4):604-615. DOI: 10.1101/gr.162172.113

[14] Shen Z, Wang H. Molecular players involved in temperature-dependent sex determination and sex differentiation in Teleost fish. Genetics, Selection, Evolution. 2014;**46**(1):26. DOI: 10.1186/1297-9686-46-26

[15] Crowder CM, Romano SN, Gorelick DA. G protein-coupled estrogen receptor is not required for sex determination or ovary function in zebrafish. Endocrinology. 2018;**159**(10):3515-3523. DOI: 10.1210/en.2018-00685

[16] Wu GC, Chang CF. Primary males guide the femaleness through the regulation of testicular *dmrt1* and ovarian *cyp19a1a* in protandrous

black porgy. General and Comparative Endocrinology. 2018;**261**:198-202. DOI: 10.1016/j.ygcen.2017.01.033

[17] Cui Y, Wang W, Ma L, et al. New locus reveals the genetic architecture of sex reversal in the Chinese tongue sole (*Cynoglossus semilaevis*). Heredity. 2018;**121**(4):319-326. DOI: 10.1038/s41437-018-0126-6

[18] Liu H, Todd EV, Lokman PM, et al. Sexual plasticity: A fishy tale. Molecular Reproduction and Development. 2017;**84**(2):171-194. DOI: 10.1002/mrd.22691

[19] Miura I. Sex determination and sex chromosomes in amphibia. Sexual Development. 2017;**11**(5-6):298-306. DOI: 10.1159/000485270

[20] Yoshimoto S, Ito M. A ZZ/ZW-type sex determination in *Xenopus laevis*. FEBS Journal. 2011;**278**(7):1020-1026. DOI: 10.1111/j.1742-4658.2011.08031.x

[21] Piprek RP, Damulewicz M, Kloc M, Kubiak JZ. Transcriptome analysis identifies genes involved in sex determination and development of *Xenopus laevis* gonads. Differentiation. 2018;**100**:46-56. DOI: 10.1016/j.diff.2018.02.004

[22] Flament S. Sex reversal in amphibians. Sexual Development. 2016;**10**(5-6):267-278. DOI: 10.1159/000448797

[23] Nakamura M. Is a sex-determining gene(s) necessary for sex-determination in amphibians? Steroid hormones may be the key factor. Sexual Development. 2013;**7**(1-3):104-114. DOI: 10.1159/000339661

[24] Oike A, Kodama M, Yasumasu S, et al. Participation of androgen and its receptor in sex determination of an amphibian species. PLoS One. 2017;**12**(6):e0178067. DOI: 10.1371/journal.pone.0178067

[25] Merchant-Larios H, Díaz-Hernández V. Environmental sex determination mechanisms in reptiles. Sexual Development. 2013;**7**(1-3):95-103. DOI: 10.1159/000341936

[26] Valenzuela N, Badenhorst D, Montiel EE, Literman R. Molecular cytogenetic search for cryptic sex chromosomes in painted turtles *Chrysemys picta*. Cytogenetic and Genome Research. 2014;**144**(1):39-46. DOI: 10.1159/000366076

[27] Pieau C, Dorizzi M, Richard-Mercier N. Temperature-dependent sex determination and gonadal differentiation in reptiles. In: Scherer G, Schmid M, editors. Genes and Mechanisms in Vertebrate Sex Determination. Switzerland: Birkhäuser Verlag Basel; 2001. pp. 117-141

[28] Georges A, Holleley CE. How does temperature determine sex? Science. 2018;**360**(6389):601-602. DOI: 10.1126/science.aat5993

[29] Janes DE, Organ CL, Stiglec R, et al. Molecular evolution of *dmrt1* accompanies change of sex-determining mechanisms in reptilia. Biology Letters. 2014;**10**(12):20140809. DOI: 10.1098/rsbl.2014.0809

[30] Rovatsos M, Augstenová B, Altmanová M, et al. Triploid colubrid snake provides insight into the mechanism of sex determination in advanced snakes. Sexual Development. 2018;**12**:251-255. DOI: 10.1159/000490124

[31] Holleley CE, O'Meally D, Sarre SD, et al. Sex reversal triggers the rapid transition from genetic to temperature-dependent sex. Nature. 2015;**523**(7558):79-82. DOI: 10.1038/nature14574

[32] Schmid M, Smith J, Burt DW, et al. Third report on chicken genes and chromosomes 2015. Cytogenetic and

Genome Research. 2015;**145**(2):78-179. DOI: 10.1159/000430927

[33] Kuroiwa A. Sex-determining mechanism in avians. Advances in Experimental Medicine and Biology. 2017;**1001**:19-31. DOI: 10.1007/978-981-10-3975-1_2

[34] Hirst CE, Major AT, Smith CA. Sex determination and gonadal sex differentiation in the chicken model. The International Journal of Developmental Biology. 2018;**62**(1-3):153-166. DOI: 10.1387/ijdb.170319cs

[35] Caetano LC, Gennaro FG, Coelho K, et al. Differential expression of the MHM region and of sex-determining-related genes during gonadal development in chicken embryos. Genetics and Molecular Research. 2014;**13**(1):838-849. DOI: 10.4238/2014. February.13.2

[36] Sánchez L, Chaouiya C. Logical modelling uncovers developmental constraints for primary sex determination of chicken gonads. Journal of The Royal Society Interface. 2018;**15**(142):20180165. DOI: 10.1098/rsif.2018.0165

[37] Elbrecht A, Smith RG. Aromatase enzyme activity and sex determination in chickens. Science. 1992;**255**(5043):467-470. DOI: 10.1126/science.1734525

[38] Vaillant S, Guémené D, Dorizzi M, et al. Degree of sex reversal as related to plasma steroid levels in genetic female chickens (*Gallus domesticus*) treated with Fadrozole. Molecular Reproduction and Development. 2003;**65**(4):420-428. DOI: 10.1002/mrd.10318

[39] Wang J, Gong Y. Transcription of CYP19A1 is directly regulated by SF-1 in the theca cells of ovary follicles in chicken. General and Comparative Endocrinology. 2017;**247**:1-7. DOI: 10.1016/j.ygcen.2017.03.013

[40] Trukhina AV, Lukina NA, Smirnov AF. Hormonal sex inversion and some aspects of its genetic determination in chicken. Russian Journal of Genetics. 2018;**54**(9):1069-1077. DOI: 10.1134/S1022795418090144

[41] Morris KR, Hirst CE, Major AT, et al. Gonadal and endocrine analysis of a gynandromorphic chicken. Endocrinology. 2018;**159**(10):3492-3502. DOI: 10.1210/en.2018-00553

[42] Ogino Y, Tohyama S, Kohno S, et al. Functional distinctions associated with the diversity of sex steroid hormone receptors ESR and AR. The Journal of Steroid Biochemistry and Molecular Biology. 2018;**184**:38-46. DOI: 10.1016/j.jsbmb.2018.06.002

[43] Weber C, Capel B. Sex reversal. Current Biology. 2018;**28**:R1234-R1236

Chapter 4

Specific Features of Sex Determination in Birds on the Example of *Gallus gallus* domesticus

Aleksandr Fedorovich Smirnov
and Antonina Vladimirovna Trukhina

Abstract

The chapter is devoted to the consideration of sex determination in birds. The appearance of heteromorphic sex chromosomes (birds) has led to the emergence of some specific W chromosomal signal, which provides estrogen control of the development of a heterogametic sex. At present, two hypotheses about sex determination in birds compete. One of these hypotheses considers the number of Z chromosomes as a key sex-determining factor, while the other hypothesis supposes the presence in W chromosome of the key gene controlling ovarian development or suppressing the appearance of testes. Into the modern scheme of the genetic control of sex determination in birds (practically within the hypothesis of dose compensation), an epigenetic mechanism was added. The appearance of gonads in birds is most likely determined by sex hormones and to the greatest extent by estrogen under the control of W chromosome. It is desirable to pay attention to noncoding RNAs, their connection with the W chromosome and their role in bird sex determination.

Keywords: sex determination, sex hormones, sex chromosomes, sex-determining genes, bird, Gallus gallus domesticus

1. Introduction

Sex is characterized by a set of features that ensure sexual reproduction. We distinguish the primary definition of sex—the emergence of one of two types of gonads (organs somatic of nature), their sexual differentiation into final system, and the development of two types germ cells. In different groups of vertebrates, different mechanisms of sex determination are realized. We consider hypothetical schemes of such a process in birds using the example of *Gallus gallus domesticus* [1]. On the one hand, chicken is an important model object of fundamental genetics, especially embryogenetics [2], and on the other hand, it has significant practical importance for humans: 210 million tons of meat and 1482 billion eggs per year [2, 3]. Both males and females are fattened in broiler production. There is currently no economically worthwhile use of the male of egg breeds. Therefore, the 1-day cockerels are destroyed, and this applies to 330 million chickens annually in the European Union alone [1, 4, 5]. This raises ethical issues, and understanding the principles of gender genetics, as well as gender selection algorithms in early embryos, is extremely important from a fundamental and applied point of view.

2. Sex determination in bird

In birds, females are heterogametic. Embryos with two Z chromosomes in birds develop as males, and those with ZW chromosomes develop as females (**Figure 1**).

At present, two hypotheses about sex determination in birds compete. One of these hypotheses considers the number of Z chromosomes as a key sex-determining factor, while the other hypothesis supposes the presence in W chromosome of the key gene controlling ovarian development or suppressing the appearance of testes. A decrease in the expression of doublesex and mab-3-related transcription factor 1 (DMRT1) protein in ovo as a result of RNA interference of this gene leads to the feminization of embryonic gonads in genetic males. In the feminized left gonad, a decrease in the expression of the SRY-box transcription factor 9 (SOX9) gene and an increase in the expression of the cytochrome P450 aromatase (CYP19A1) gene were observed. This observation confirms the hypothesis of the presence of dose compensation in the DMRT1 gene. In feminized right gonad, the expression of the corresponding genes is very different, indicating differential sensitivity to DMRT1 between the left and right gonads. Germ cells in feminized gonads are distributed as in the ovaries. All this indicates that the DMRT1 gene is necessary for the development of testes [6]. But there is no clear evidence that this gene controls the primary sex determination (the appearance of testis or ovary).

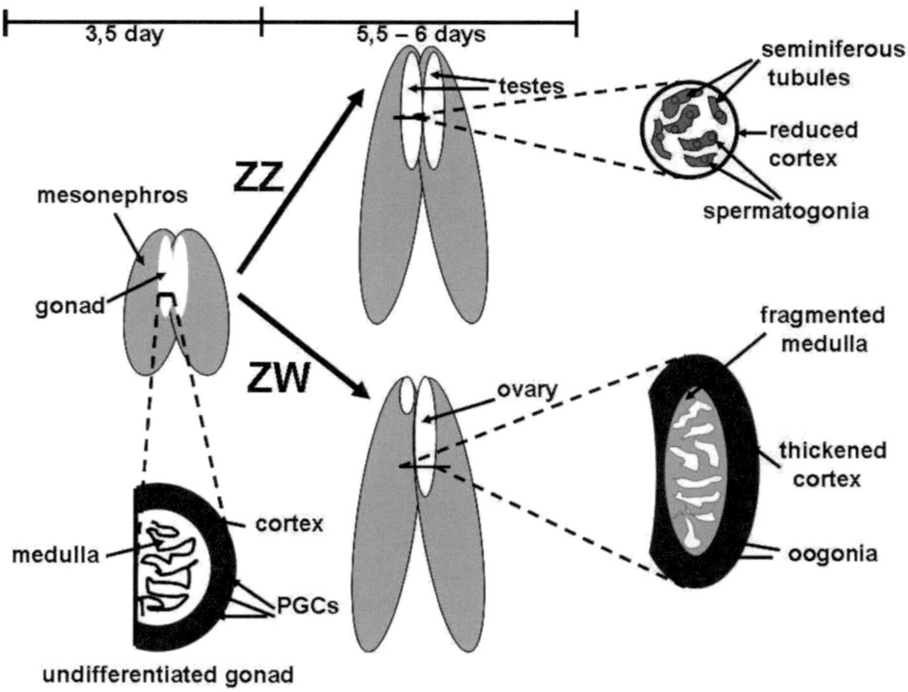

Figure 1.
A hypothetical scheme of primary sex differentiation in Gallus gallus. Initially gonads are undifferentiated. They consist of an outer cortex and underlying medullary layer. Primordial germ cells (PGCs) are visible on 3.5 day and are located mainly in the cortex. On the 5.5–6th day (stage 28–30), bilateral testes appear in ZZ embryos, and in ZW embryos, the left gonad gives ovary and the right one regresses (adapted from [1]).

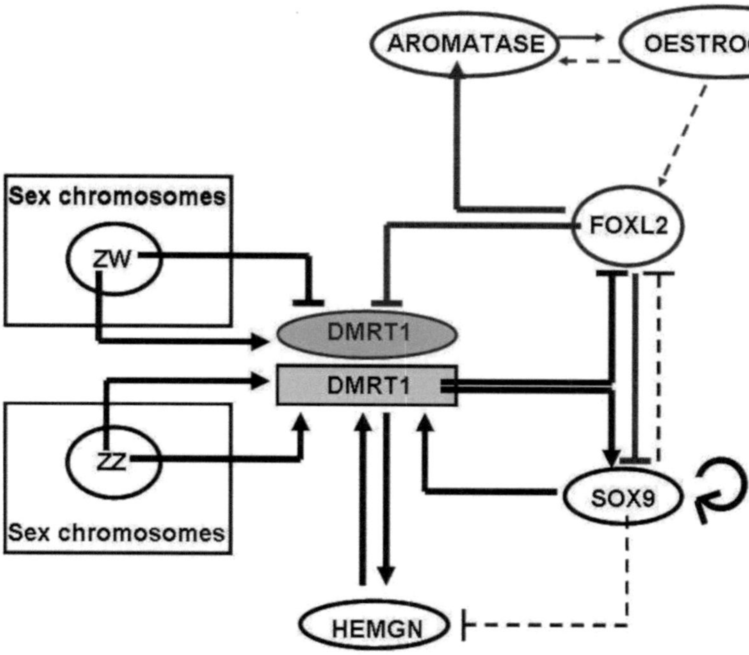

Figure 2.
The simplified gene regulatory network controlling primary sex determination of chicken. Black (from male genes) and gray (from female genes) arrows represent positive (→) and negative (⊣) interactions, and dashed arrows indicate indirect or proposed interactions between Z and Z chromosomes and between Z and W chromosomes (adapted from [9]).

Figure 2 presents a hypothetical scheme of genetic control of primary sex differentiation in Gallus gallus. The gonad appears on the 3.5th day (stage 22) as thickening on the surface of mesonephros. It consists of the epithelial layer of somatic and germ cells and medullary cordate layer (epithelial cords), which is mixed with mesenchymal cells. On the 6.5th day (stage 30), the first genes that determine sex are activated.

Into the modern scheme of the genetic control of sex determination in birds (practically within the hypothesis of dose compensation), an epigenetic mechanism was added ([7, 8], **Figure 2**).

3. Sex reversal

Sex reversal is the phenomenon whereby organisms developing at sex-specific conditions hatch the opposite sex. This can be caused by factors acting as estrogen promoters or inhibitors, increasing or decreasing the number of female offspring, through controlling aromatase [10]. Synthetic aromatase inhibitors (an enzyme catalyzing the synthesis of estrogens) can cause steady inversion of sex from female to male. In this case, the left gonad becomes an ovotestis, or a testis, and the right gonad becomes a testis. In most experiments injection of aromatase inhibitors in ovo is carried out on the 3rd or 4th day of incubation [11, 12]. The proportion of individuals with sex inversion and two testes is increased with an earlier introduction of the inhibitor at the beginning of incubation [13]. However, females with inversed sex have got testes with an abnormal development and with an abundance of abnormal spermatozoa in the seminiferous tubules. The experimental results suggest that the abnormal development of the testes in the sex-reversed female chicken is jointly regulated by sex-related genes and long noncoding RNAs (lncRNA); Wnt (the term **wnt** is an amalgam of **wingless** (Wg) and **int**) and transforming growth factor beta/the bone morphogenetic protein signaling pathways (TGFβ/BMP signaling pathways) play an essential role in regulating developmental pathways during embryogenesis, including a very important role in the differentiation of gonads and in maintaining their function in chickens [14]. Unfortunately at the moment there is no clear concrete data on these issues. It was previously noted that the earliest expression of aromatase in birds is detected only on the 5th day of embryonic development in the medullary layer of the ovary. It is also worth noting that the appearance of aromatase has recently been demonstrated already in the mother's body, after oogenesis in the theca of early follicles [15]. The effect of estrogens on female gonadogenesis follows from the classical scheme of primary sex determination in Gallus gallus. At the same time, in experimental males injection of estradiol results in reversible feminization [16].

In recent study Morris et al. described the gonads and endocrine profile of a gynandromorphic chicken. Its right side had male features, and the left side had female features. Almost all cells (96%) on the right side had a ZZ karyotype, and the left side had a mixture of cells with ZZ and ZW karyotypes. Moreover, the number of cells with the ZW karyotype was much smaller than with the ZZ karyotype. A reduced percentage of cells with the ZW karyotype (23%) did not affect the manifestation of female traits. Based on these observations, Morris et al. concluded that even a small percentage of ZW cells is sufficient for female differentiation. They also confirm the hypothesis of the existence of cell autonomy, on which sexual differentiation in birds depends [17].

Thus, comparing the results of experiments on the inversion of sex and the search for a sex-determining gene in chickens, it can be assumed that the primary determination of the sex of birds is determined by the estrogens content in the early stages of embryogenesis. The role of DMRT1 is associated with the correct regular inclusion of spermatogenesis genes.

4. Female heterogametic system

A heterogametic sex is a genetically determined sex that corresponds to the presence in the cells of the body of two different sex chromosomes or one, in a double dose leading to the formation of an alternative sex. Heterogametic individuals give two groups of gametes (according to the content of different sex chromosomes). In animals with a heterogametic male sex, the letters X and Y are used to designate sex chromosomes. Individuals, normally carrying a pair of sex chromosomes X and Y or one chromosome X, are males, and two chromosomes X are females. This group includes mammals, most species of insects, and many other groups. If the heterogamous sex is female, then other designations for sex chromosomes are used—Z and W. Individuals with the genotype ZZ are males; ZW are females. Heterogamous female sex is characteristic of Lepidoptera insects, reptiles, and birds. In these species, sex is not determined during fertilization, but during meiosis. There is no full dose compensation. Female heterogametic systems are very different from the male heterogametic systems. For example, dosage compensation is incomplete in birds. Z-linked genes are higher expressed in males (with two Z chromosomes) than in females (with one Z chromosome), and only a subgroup of genes have the same dose of expression between the sexes. Two male hypermethylated (MHM) regions (MHM1 and MHM2) were found on the Z chromosome with extreme differences in DNA

	Female heterogametic system	Male heterogametic system
System of sex chromosomes	ZW/ZZ	XX/XY
Gender carrying with two types of gametes	Female	Male
Type of gametes that determines gender	Eggs	Spermatozoids
An event that leads to sex determination	Meiosis (first division)	Fertilization
Moment of sex determination regarding fertilization	Long before fertilization	In the moment of fertilization
Type of gametes in which material is stored for the development of a heterogametic sex	Eggs	Eggs
Role of eggs before fertilization	Storage of substances necessary for the development of zygotes, including various types of RNA, during egg maturation	Storage of substances necessary for the development of zygotes, including various types of RNA, during egg maturation
Role of sperm in egg fertilization	Start of development, introducing genetic material associated with the Z chromosome	Start of development, introducing genetic material associated with the Y chromosome
Possible involvement of sex chromosomes in the storage of substances	Possible involvement of Z or W chromosomes	Only the X chromosome is involved
Dose compensation	Incomplete for Z (male)	Complete for X (female)
Feature of the alternative sex chromosome (W or Y)	Almost completely consists of repeats, completely heterochromatic	Almost completely consists of repeats, completely heterochromatic
Origin of alternative sex chromosome (W or Y)	Maternal	Paternal

Table 1.
The similarity and difference between the female and male heterogametic system.

methylation between male and female chicken [18, 19]. Most of the samples analyzed showed a limited effect of MHM1 on transcription of DMRT1. The role of DMRT1 in determining sex in chickens, which may depend on developmental stage and tissue, needs further evaluation [19]. The similarity and difference between the female and male heterogametic system are shown in **Table 1**.

The genome can differentiate tissue-specific. An excellent example of this is the presence of the germline-restricted chromosome (GRC) in the genome of some songbirds. It is absent in somatic cells. GRC is inherited through the female germline and eliminated from the somatic cells during early embryogenesis. Also it was found that GRC contains genes that are paralogous to genes on autosomes and the Z chromosome. More than 38 GRC-linked genes were found in a large number of copies. There are no mobile elements on GRC. The GRC is enriched in genes that are highly expressed in gonads and are involved in the development of female gonads. Germline-restricted chromosomes are probably widespread in the highly dynamic evolutionary history of songbirds, which leads to significant differences between the genome of germline cells and the genome of somatic cells. This is a new mechanism for minimizing the genetic conflict between germ line cells and somatic cells [20, 21].

5. W chromosome and its role

Chicken W chromosome has a length of about 7.08 Mb and contains 28 genes. It was shown that the decay of genes specific for the W chromosome is not random, and therefore it was suggested that the content of the surviving genes associated with the W chromosome was under high evolutionary pressure due to absence of recombination. There are no genes on the W chromosome that determine sex. In the evolution of bird karyotypes, the W chromosome has undergone extensive degradation and accumulation of repetitive DNA [22, 23]. Notably, the most common W chromosome-specific repeats *XhoI*, *EcoRI*, *SspI*, and newly described (GGAAA) n are compacted in densely packed chromomeres and do not transcribe at the lampbrush stage. *XhoI*, *EcoRI*, and *SspI* were shown to lack any significant homology between Galliformes, Ciconiiformes, and Passeriformes. Komissarov et al.

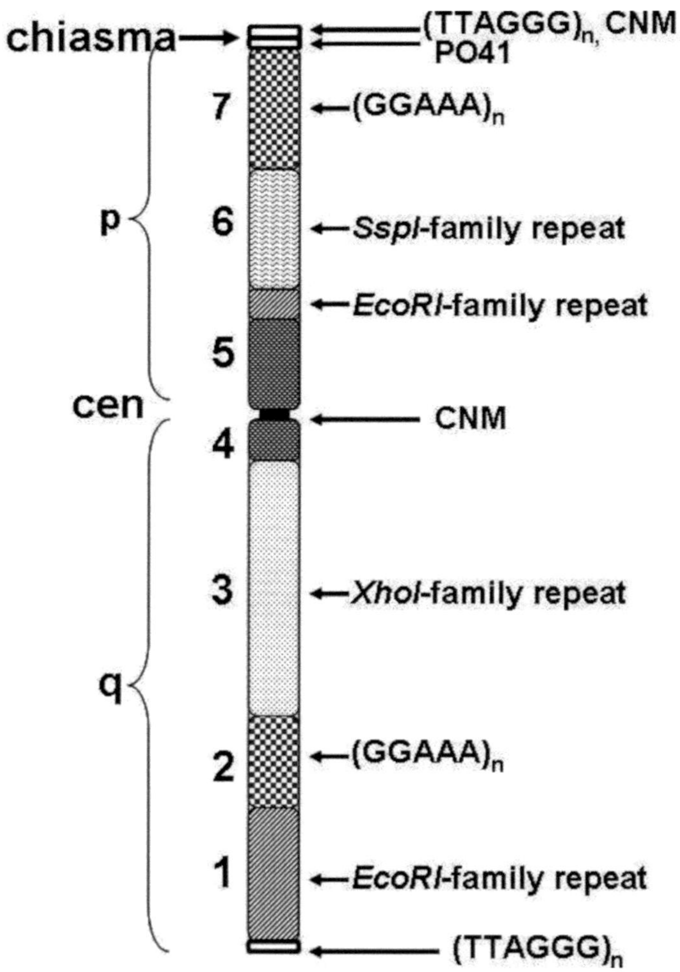

Figure 3.
Chicken W lampbrush chromosome diagram with localized repeat blocks. Chromomeres are numbered from the free end of the W chromosome to the chiasma region. Loop-related repeats are indicated by thin arrows. Non-transcribed repeat locations are marked with thick arrows (adapted from [23]).

Ploidy	Sex chromosomes	*ZUF*	Z dosage	Z/A ratio	Phenotype
Normal					
2A	ZZ	—	1*2Z = 2	1,00	M
2A	ZW	+	2*1Z = 2	1,00	F
Aneuploid					
2A	ZO	—	1*1Z = 1	0,50	M (dead)
2A	ZZZ	—	1*3Z = 3	1,50	M (dead)
2A	ZZW	+	2*2Z = 4	2,00	F (dead)
2A	ZWW	+	2*1Z = 2	1,00	F
Triploid					
3A	ZZZ	—	1*3Z = 3	1,00	M
3A	ZZW	+	2*2Z = 4	1,33	F/M
3A	ZWW	+	2*1Z = 2	0,67	F (dead)

ZUF = Z upregulated factor.

Table 2.
The number of Z chromosomes and phenotype of aneuploid chickens [27].

suggested independent accumulation of specific DNA repeats, which occurred after initial divergence of Z and W chromosomes (**Figure 3**, [23]).

The determination of the function of repeating sequences on Y or W chromosomes is in its infancy, but it is clear that these sequences play a functional role in gene regulation and chromatin structure [23]. Banded krait minor (Bkm) satellite repeat in the W chromosome consists of tandem arrays of GATA nucleotides. Tissue-specific protein that binds specifically to Bkm repeats, known as Bkm-binding protein (BBP), is involved in the coordinated decondensation of the heterogametic sex chromosomes in germ cells. It is known that GATA repeats play a conservative role of insulators [24, 25].

The transcriptome of sex-inversed chickens has a wide variety of lncRNA classes compared to the transcriptome of ordinary males. lncRNAs are classified as a separate class of non-protein-coding genes. These genes do not encode proteins and do not have open reading frames; their functional constraints differ from those for protein-encoding genes or genes where most of the nucleotide sequence is necessary for a function similar with tRNA. Despite this, relatively few lncRNAs have been studied in detail. They present high sequence divergence between species. Nonetheless, similar or equivalent lncRNAs perform the same functions in different organisms [26]. The W chromosome is necessary for the appearance of the ovary. Individuals containing it are always females (**Table 2**) [27, 28]. A specific factor, called F or ZUF (Z upregulated factor), is involved in gene dose compensation in determining sex. This W-specific factor can control gene expression from the single Z chromosome of the heterogametic sex [27]. Unlike mammals, we never see birds with differences in the number of Z and W chromosomes; it seems that there are no bird equivalent to women with XO with one X chromosome and men with XXY chromosomes. Perhaps such changes are lethal in birds.

6. The ability to choose the sex of the descendants of female birds

There is some experimental evidence that sexual cleavage in birds may not be random [29, 30]. This corresponds to the previously reported effects of hormonal manipulations on the offspring. It appears that there is a critical level of corticosterone needed to manipulate the sex of the offspring and that this level must be achieved within a relatively short time (**Figure 4**, [31–34]).

Proposed preovulatory mechanisms for a primary sex ratio bias in birds include (1) asynchronous sex-specific follicular development, (2) segregation distortion or meiotic drive, and (3) selective resorption of postmeiotic and preovulatory follicle (**Figure 5**).

There are hypotheses that explain sex ratio bias (SR bias). These are sex-specific fertilization and sex-specific embryo mortality. Now they are not relevant. Hormonal fluctuations caused by external factors that the female encounters alter the normal process of meiosis I (MI, the first meiotic division), blocking the segregation of the first polar body (PB) (**Figure 6**, [32]). But not only maternal hormones and other components of the yolk might affect the result. Temperature is another factor responsible for differential embryonic mortality in birds.

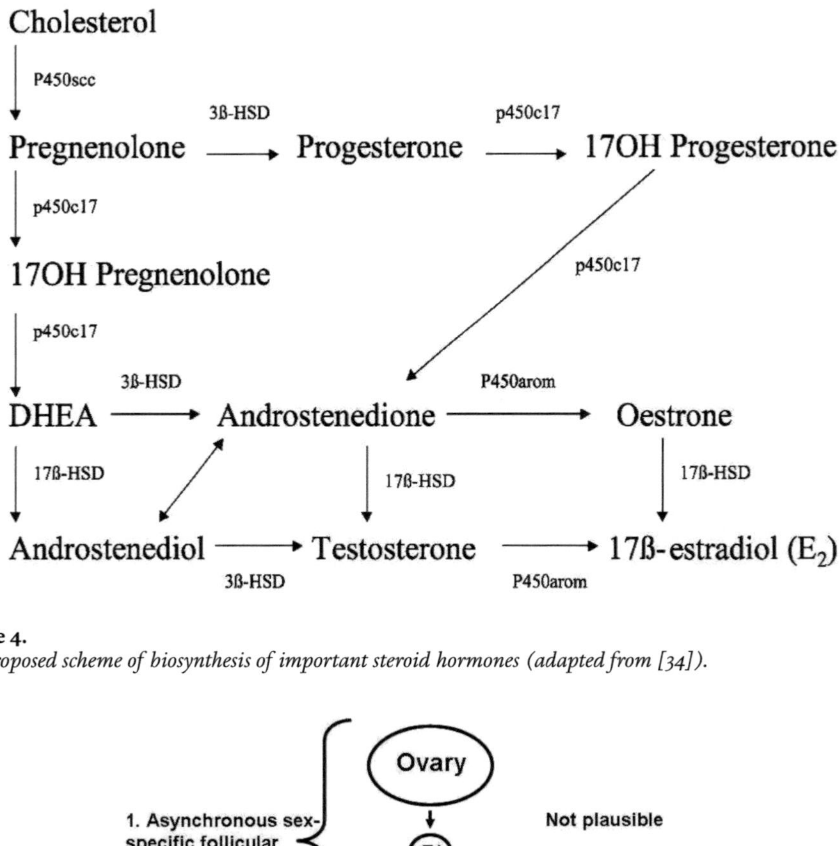

Figure 4.
The proposed scheme of biosynthesis of important steroid hormones (adapted from [34]).

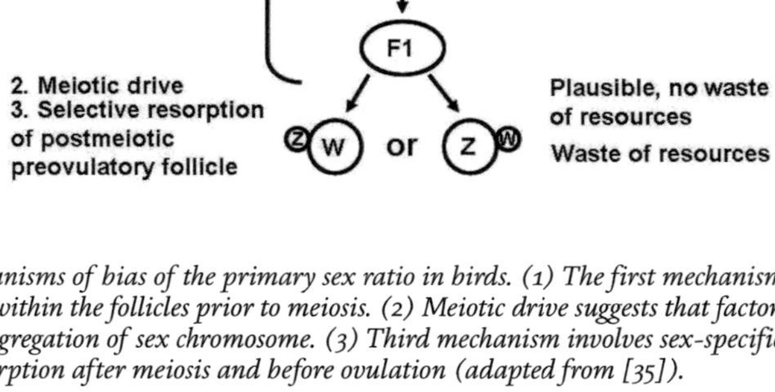

Figure 5.
The suggested mechanisms of bias of the primary sex ratio in birds. (1) The first mechanism implies the existence of factors within the follicles prior to meiosis. (2) Meiotic drive suggests that factors during meiosis I cause nonrandom segregation of sex chromosome. (3) Third mechanism involves sex-specific follicle abortion and subsequent resorption after meiosis and before ovulation (adapted from [35]).

Gonadal sex determination refers to the decision to differentiate as a testis or an ovary [29], such that sex determination occurs chronologically before sex differentiation and glucocorticoids (GCs) could interfere with both processes. For instance, in avian species, corticosterone may influence sex chromosome segregation at the first meiotic division, thereby acting directly on sex determination.

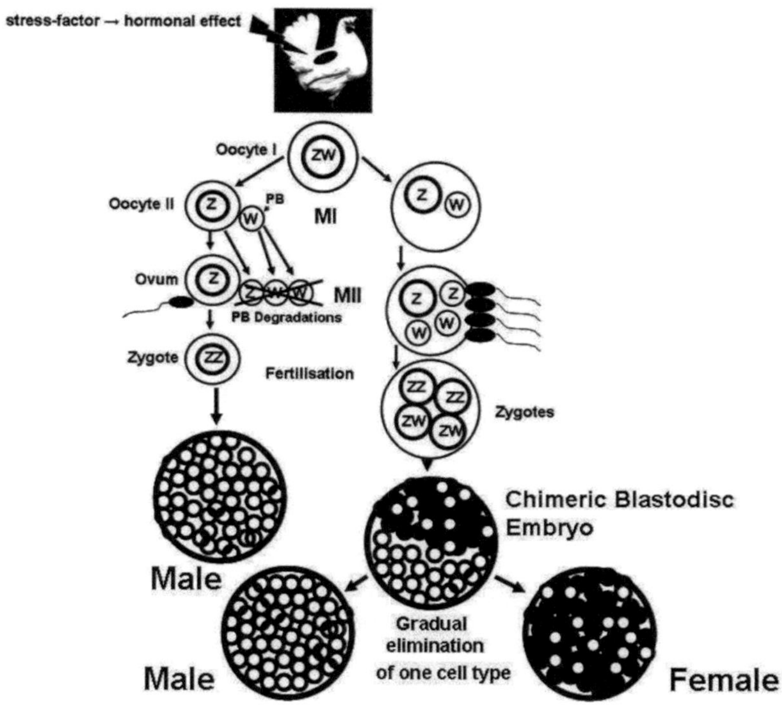

Figure 6.
Exclusion of the first polar body (PB) as a possible mechanism of deviation from the primary SR bias in birds (adapted from [32]).

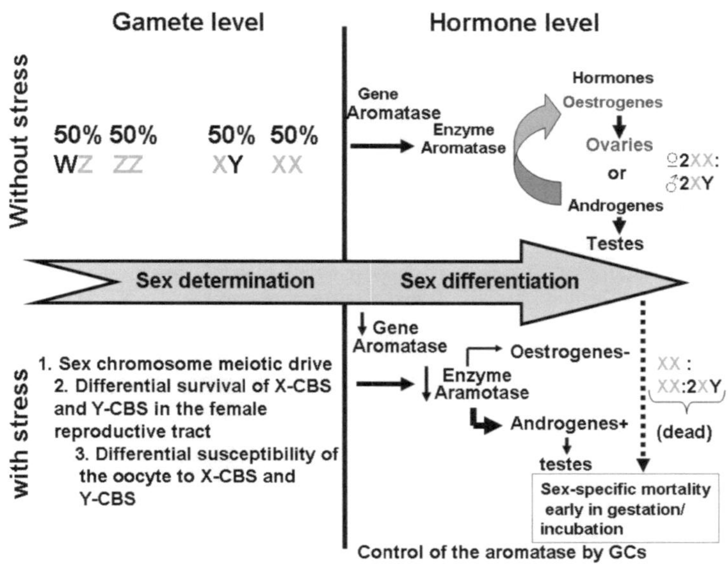

Figure 7.
Mechanism of influence of glucocorticoids (GCs) on determination and differentiation of sex in vertebrates (adapted from [10]).

This is possible because, in birds, females are the heterogametic sex (ZW chromosomes) and have therefore a high degree of control over the sex ratio of the offspring they produce (**Figure 7**, [10, 34]). Future studies regarding sex ratio bias in birds should focus on the cellular and molecular mechanisms of sex ratio bias by examining the gene and protein expression during meiosis using genomics and proteomics techniques.

7. Conclusion

For birds, the DMRT1 gene is considered as the sex-determining gene. The main argument is the change in the sex of male after turning it off with the use of interfering RNA on the 1st day of incubation [6]. There is also the possibility of changing the sex by introducing an aromatase inhibitor, usually on day 5.5 or even at the very beginning of incubation in the direction of the female → male [11–13]. According to recent data, aromatase appears already in the layer of early follicles [14]. The unstable alteration of males into females after injection of estrogen is worth mentioning [16]. The indispensable presence of the W chromosome in females suggests that synthesis of estrogens is induced by genes located on this chromosome. It can be assumed that this process is triggered by the inclusion of a number of factors using lncRNA. The primary sex determination is the onset of testis or ovary before differentiation (**Figure 8**).

Dmrt1 is unusual in that it is expressed by both Sertoli cells and germ cells. Deletion or inactivation of DMRT1 gene in human resulted in XY male-to-female sex reversal. DMRT1 is required for a stable testis phenotype. It is well known that retinoic acid (RA) signaling between Sertoli cells and germ cells is essential for adult mammalian spermatogenesis. In the absence of DMRT1, RA signaling may also activate genes that can drive male-to-female transdifferentiation. Gonad sex identity may be able to switch at adult stage in some fish species. In mammals the gonadal phenotype remains stable for the rest of life. In birds the function of DMRT1 is not primary sex determination, but maintenance of already formed testis [37]. The

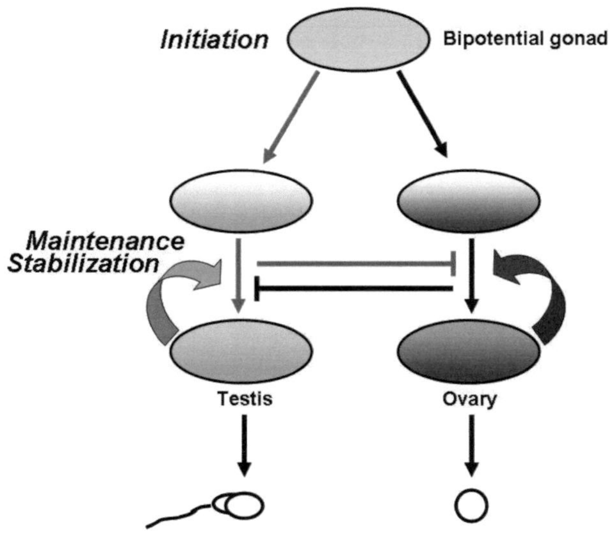

Figure 8.
Stages of gonadal sex determination: the stage of initiation, maintenance, and stabilization. The sex reversal can only occur at stages of initiation and maintenance. The complete sex alteration can only occur at the stabilization stage (adapted from [36]).

appearance of gonads in birds is most likely determined by sex hormones and to the greatest extent by estrogen under the control of W chromosome. The role of the DMRT1 gene is to maintain and develop testis.

It can be assumed that the W chromosome causes the appearance of ovaries through the control of estrogen synthesis and their interaction with the corresponding receptors. The W chromosome seems to have no specifically female genes. Over the years, it was believed that the main function of RNA is to act as an intermediary in the process of reading a protein sequence from the gene encoding it. Therefore, one of the biggest surprises in modern biology was the discovery that protein-coding sequences comprise less than 2% of the total genome; then it has been found that at least 90% of the human genome is actively transcribed. It should be noted that lncRNA are transcripts that do not encode protein and have a length of more than 200 nucleotides. However, despite debates, the proof that certain lncRNA clearly play first-line roles in development, pluripotency, dosage compensation, establishment of chromatin structure, genome maintenance, and as tumor suppressors or oncogenes is not contested [38, 39]. It assumes that the development of testis in male of chicken is co-regulated by sex-related genes and long noncoding RNA, and Wnt and TGFβ/BMP signaling pathways have a very important role in gonadal differentiation. It is desirable to pay attention to noncoding RNAs, their connection with the W chromosome and their role in bird sex determination.

Acknowledgements

This research was supported by a grant 17-04-01321A Russian Foundation for Basic Research (RFBR).

Conflict of interest

The authors declare that the research was conducted in the absence of any commercial or financial relationships that could be considered as a potential conflict of interest.

Author details

Aleksandr Fedorovich Smirnov and Antonina Vladimirovna Trukhina*
Saint-Petersburg State University, Saint-Petersburg, Russia

*Address all correspondence to: a.trukhina@spbu.ru

References

[1] Smirnov AF, Trukhina AV. Molecular-Genetic Mechanisms of Sex Determination in Animals. Irvine, CA, United States: Scientific Research Publishing Inc.; 2017. p. 126. ISBN: 978-1-61896-390-1

[2] Gammill LS, Jacques-Fricke B, Roffers-Agarwal J. Embryological and genetic manipulation of chick development. In: Pelegri F, editor. Vertebrate Embryogenesis. Methods in Molecular Biology. Vol. 1920. New York, NY: Humana; 2019

[3] Giersberg MF, Kemper N. Rearing male layer chickens: A German perspective. Agriculture. 2018;**8**:176. DOI: 10.3390/agriculture8110176

[4] Krautwald-Junghanns ME, Cramer K, Fischer B, et al. Current approaches to avoid the culling of day-old male chicks in the layer industry, with special reference to spectroscopic methods. Poultry Science. 2018;**97**(3):749-757. DOI: 10.3382/ps/pex389

[5] Reithmayer C, Mußhoff O. Consumer preferences for alternatives to chick culling in Germany. Poultry Science. 2019;**98**(10):4539-4548. DOI: 10.3382/ps/pez272

[6] Smith CA, Roeszler KN, Ohnesorg T, et al. The avian Z-linked gene DMRT1 is required for male sex determination in the chicken. Nature. 2009;**461**(7261):267-271. DOI: 10.1038/nature08298

[7] Kuroiwa A. Sex-determining mechanism in avians. Advances in Experimental Medicine and Biology. 2017;**1001**:19-31. DOI: 10.1007/978-981-10-3975-1_2

[8] Hirst CE, Major AT, Smith CA. Sex determination and gonadal sex differentiation in the chicken model. The International Journal of Developmental Biology. 2018;**62**(1-2-3): 153-166. DOI: 10.1387/ijdb.170319cs

[9] Sánchez L, Chaouiya C. Logical modelling uncovers developmental constraints for primary sex determination of chicken gonads. Journal of the Royal Society Interface. 2018;**15**(142). pii: 20180165. DOI: 10.1098/rsif.2018.0165

[10] Geffroy B, Douhard M. The adaptive sex in stressful environments. Trends in Ecology & Evolution. 2019;**34**(7):628-640. DOI: 10.1016/j.tree.2019.02.012

[11] Elbrecht A, Smith RG. Aromatase enzyme activity and sex determination in chickens. Science. 1992;**255**(5043):467-470. DOI: 10.1126/science.1734525

[12] Vaillant S, Guémené D, Dorizzi M, et al. Degree of sex reversal as related to plasma steroid levels in genetic female chickens (*Gallus domesticus*) treated with Fadrozole. Molecular Reproduction and Development. 2003;**65**(4):420-428. DOI: 10.1002/mrd.10318

[13] Trukhina AV, Lukina NA, Smirnov AF. Hormonal sex inversion and some aspects of its genetic determination in chicken. Russian Journal of Genetics. 2018;**54**(9):1069-1077. DOI: 10.1134/S1022795418090144

[14] Li D, Deng J, Sun T, et al. Transcriptomic profile of testis in adult sex-reversed female chicken. Molecular Biology. 2020. (in press)

[15] Wang J, Gong Y. Transcription of CYP19A1 is directly regulated by SF-1 in the theca cells of ovary follicles in chicken. General and Comparative Endocrinology. 2017;**247**:1-7. DOI: 10.1016/j.ygcen.2017.03.013

[16] Tagirov MT. Sex determination and control mechanisms in birds.

Biotechnologia Acta. 2013;**6**(1):62-72. DOI: 10.15407/biotech6.01.062

[17] Morris KR, Hirst CE, Major AT, et al. Gonadal and endocrine analysis of a gynandromorphic chicken. Endocrinology. 2018;**159**(10):3492-3502. DOI: 10.1210/en.2018-00553

[18] Teranishi M, Shimada Y, Hori T, et al. Transcripts of the MHM region on the chicken Z chromosome accumulate as non-coding RNA in the nucleus of female cells adjacent to the DMRT1 locus. Chromosome Research. 2001;**9**(2):147-165. DOI: 10.1023/A:1009235120741

[19] Sun D, Maney DL, Layman TS, et al. Regional epigenetic differentiation of the Z chromosome between sexes in a female heterogametic system. Genome Research. 2019;**29**(10):1673-1684. DOI: 10.1101/gr.248641.119

[20] Kinsella CM, Ruiz-Ruano FJ, Dion-Côté A-M, et al. Programmed DNA elimination of germline development genes in songbirds. Nature Communications. 2019;**10**(5468):1-10. DOI: 10.1101/444364

[21] Torgasheva AA, Malinovskaya LP, Zadesenets KS, et al. Germline-restricted chromosome (GRC) is widespread among songbirds. Proceedings of the National Academy of Sciences of the United States of America. 2019;**116**(24):11845-11850. DOI: 10.1073/pnas.1817373116

[22] Bellott DW, Skaletsky H, Cho TJ, et al. Avian W and mammalian Y chromosomes convergently retained dosage-sensitive regulators. Nature Genetics. 2017;**49**(3):387-394. DOI: 10.1038/ng.3778

[23] Komissarov AS, Galkina SA, Koshel EI, et al. New high copy tandem repeat in the content of the chicken W chromosome. Chromosoma. 2018;**127**(1):73-83. DOI: 10.1007/s00412-017-0646-5

[24] Ezaz T, Deakin JE. Repetitive sequence and sex chromosome evolution in vertebrates. Advances in Evolutionary Biology. 2014;**2014**. Article ID: 104683. 9 p. DOI: 10.1155/2014/104683

[25] Singh L, Wadhwa R, Naidu S, Nagaraj R, Ganesan M. Sex- and tissue-specific Bkm(GATA)-binding protein in the germ cells of heterogametic sex. The Journal of Biological Chemistry. 1994;**269**(41):25321-25327

[26] Murillo-Maldonado JM, Riesgo-Escovar JR. The various and shared roles of lncRNAs during development. Developmental Dynamics. 2019;**248**(11):1059-1069. DOI: 10.1002/dvdy.108

[27] Graves JA. Sex and death in birds: A model of dosage compensation that predicts lethality of sex chromosome aneuploids. Cytogenetic and Genome Research. 2003;**101**(3-4):278-282. DOI: 10.1159/000074349

[28] Küpper C, Augustin J, Edwards S, et al. Triploid plover female provides support for a role of the W chromosome in avian sex determination. Biology Letters. 2012;**8**(5):787-789. DOI: 10.1098/rsbl.2012.0329

[29] Heinsohn R, Legge S, Barry S. Extreme bias in sex allocation in Eclectus parrots. Proceedings of the Biological Sciences. 1997;**264**(1386):1325-1329. DOI: 10.1098/rspb.1997.0183

[30] Pryke SR, Griffith SC. Genetic incompatibility drives sex allocation and maternal investment in a polymorphic finch. Science. 2009;**323**(5921):1605-1607. DOI: 10.1126/science.1168928

[31] Pinson SE, Wilson JL, Navara KJ. Timing matters: Corticosterone injections 4 h before ovulation bias sex ratios towards females in chickens. Journal of Comparative Physiology.

B. 2015;**185**(5):539-546. DOI: 10.1007/s00360-015-0897-5

[32] Tagirov M, Rutkowska J. Chimeric embryos—Potential mechanism of avian offspring sex manipulation. Behavioral Ecology. 2013;**24**(4):802-805. DOI: 10.1093/beheco/art007

[33] Wrobel ER, Molina E, Khan NY, et al. Androgen and mineralocorticoid receptors are present on the germinal disc region in laying hens: Potential mediators of sex ratio adjustment in birds? General and Comparative Endocrinology. 2019;**287**:113353. DOI: 10.1016/j.ygcen.2019.113353

[34] Bruggeman V, Van AP, Decuypere E. Developmental endocrinology of the reproductive axis in the chicken embryo. Comparative Biochemistry and Physiology. Part A, Molecular & Integrative Physiology. 2002;**131**(4):839-846. DOI: 10.1016/s1095-6433(02)00022-3

[35] Aslam ML, Woelders H. Steroid hormones and female energy balance: Relation to offspring primary sex ratio. In: Hester PY, editor. Egg Innovations and Strategies for Improvements. United States: Academic Press; 2017. pp. 47-54. DOI: 10.1016/B978-0-12-800879-9.00005-6

[36] Weber C, Capel B. Sex reversal. Current Biology. 2018;**28**(21):R1234-R1236. DOI: 10.1016/j.cub.2018.09.043

[37] Huang S, Ye L, Chen H. Sex determination and maintenance: The role of DMRT1 and FOXL2. Asian Journal of Andrology. 2017;**19**(6):619-624. DOI: 10.4103/1008-682X.194420

[38] Balashenko NA, Dromashko SE. Long non-coding RNAs and their functions. Vestsi Natsyyanal'nai akademii navuk Belarusi. Seryya biyalagichnych navuk. Proceedings of the National Academy of Sciences of Belarus. 2017;**4**:110-119. (in Russian)

[39] Peng Y, Chang L, Wang Y, et al. Genome-wide differential expression of long noncoding RNAs and mRNAs in ovarian follicles of two different chicken breeds. Genomics. 2019;**111**(6):1395-1403. DOI: 10.1016/j.ygeno.2018.09.012

Section 3

Gene Expression and Regulation

Chapter 5

Transcriptional and Epigenetic Regulation of Krüppel-Like Transcription Factors

Morgan Salmon

Abstract

Krüppel-like factors (KLFs) are a family of zinc finger transcription factors (ZF-TF) that are now known to be involved in complex biological processes including cancer, proliferation, and cardiovascular disease as well as developmental processes. KLFs first gained notoriety when it became known that they are crucial for promoting and maintenance of stem cell pluripotency. Over the past 20 years since the discovery of Krüppel-like factor 1 (KLF1), this transcription factor family has grown to include 18 members and 7 closely related members of the specificity protein 1 (Sp1) family. In the present study, we review the mechanisms related to regulation of KLFs by direct promoter activation or repression. We will also review and discuss some mechanisms of posttranslational modifications that could affect KLF function. We seek to understand how these transcriptional regulators are themselves regulated and how that regulation could become aberrant during various disease processes.

Keywords: Krüppel-like zinc finger proteins, transcription, posttranslational modification, epigenetics, RNA, promoters

1. Introduction

The specificity protein 1 (Sp1)/Krüppel-like factor (KLF) proteins are a family of highly conserved transcription factors that are characterized by the presence of three highly homologous Cys2/His2-type zinc fingers near the C-terminus that bind GC/CACCC box. Amino acid sequences in the transcription activation/repression domains are less conserved among family members; however, there are subfamilies based on sequence similarities within this group. These subfamilies tend to share co-activators or co-repressors to aid in how they regulate genes. So far, seven members in the specificity protein (Sp) subgroup and 18 members in the KLF subgroup have been identified in mammalian cells [1]. This family of transcription factors is able to function as both transcriptional activators and repressors based on the gene and cellular contexts. KLFs gained notoriety as Krüppel-like factor 4 (KLF4), Krüppel-like factor 2 (KLF2), and Krüppel-like factor 5 (KLF5) were suggested to be important for embryonic stem cells and stem cell reprogramming [2–7] alongside Oct4, Sox2, and Nanog. However, we have only begun to touch the surface of the transcriptional control these factors exert during embryonic development, maintenance of normal function, and the breakdown of normal processes seen in many diseases.

The goal of this chapter is to begin to describe our current knowledge of how the KLFs are regulated during development or disease. We seek to begin to understand the ways cells either promote or repress the presence of the KLFs through a variety of transcriptional and translational mechanisms.

2. Regulation by and of KLFs

2.1 Krüppel-like factor 1

Krüppel-like factor 1 (KLF1) or erythroid Krüppel-like factor is an essential transcription factor for erythroid development and was found to be key in the regulation of many facets of blood development. KLF1 is expressed in the developing blood as well as being weakly expressed in mast cells [1]. KLF1 is key to blood development as Klf1−/− mice die around E14 due to severe anemia [8]. Several studies also showed KLF1 is able to directly bind to the β-globin promoter to activate the gene's transcription as part of fetal hematopoiesis in the liver [9, 10]. The null embryos provided a wealth of knowledge about KLF1 early on, suggesting that β-thalassemia could be linked with KLF1 deletions [11]. More recent studies have also shown that KLF1 is able to either directly or indirectly repress the transcription of the γ-globin gene to promote the expression of β-globin during blood development [12].

In humans, >140 KLF1 variants, causing different erythroid phenotypes, have been described. The KLF1 Nan variant, a single amino acid substitution (p.E339D) in the DNA-binding domain, causes hemolytic anemia and is dominant over wild-type KLF1 [13]. This variant in the developing liver demonstrates defects in erythroid maturation that resemble those seen with the KLF1−/−, again demonstrating the importance of KLF1 in blood development. Furthermore, recent studies suggest that there is an enhancer element in the KLF1 gene that is susceptible to methylation and that elevated levels of methylation in that region correlate with patients with juvenile myelomonocytic leukemia (JMML) [14]. KLF1 was also found to play a role in the inhibition of megakaryocytes while also stimulating erythroid lineages at the same time [15].

2.2 Krüppel-like factor 2

Krüppel-like factor 2 or lung Krüppel-like factor (LKLF) was isolated in humans in 1999 and found to be 85% similar in nucleotide identity and 90% similar in its amino acids to mouse and located on chromosome 19p13.1 [16]. Of special interest, a region of 75 nucleotides within its proximal promoter was found to be identical between human and mouse [16]. This identical region in the mouse and human promoters for KLF2 has been found to be critical for its regulation in lung, blood, endothelial cell, and T lymphocyte development [15–22]. KLF2 was shown to be essential for normal development within mice, and knockout embryos were lethal around day 12.5 and lung function was also severely impaired in KLF2−/− chimeras [22]. KLF2 expression appears to also be important for the maintenance of normal lung function, as methylation of KLF2 was associated with metastasis and worsening prognosis in non-small-cell lung cancer [23].

KLF2 was also shown to be essential for early erythropoiesis and regulation of the β-globin gene, and klf2−/− mice also exhibited hemorrhage in developing blood cells [17]. In mature T cells, KLF2 is required for T-cell trafficking, and elimination of KLF2 in T cells affects the expression of sphingosine-1-phosphate receptor and CD2L and beta7 integrins, receptors all important in T-cell trafficking [18, 24].

ERK5 was also shown to be important in T-cell activation, and ERK5−/− cells were unable to activate genes for T-cell function [25, 26].

KLF2 is also an important regulator of heart and aorta development and normal maintenance of endothelial cells [27–29]. KLF2 has been shown to be activated by shear stress through the conserved 75-base pair region in the human and mouse promoters [30]. This region was shown to requite PI3K for activation and PCAF (p300/CAMP-response element-binding protein-associated factor) and heterogeneous nuclear riboprotein D to induce acetylation of H3 and H4 histones [31]. Additional riboproteins and acetyltransferases such as HnRNP-U, hnRNP-D, and p300 were also found to bind via this conserved region in the KLF2 promoter [32]. KLF2 was also found to be activated by nucleolin in endothelial cells following shear stress, and activation via nucleolin was also PI3K dependent [33].

In terms of a negative regulation of KLF2 in endothelial cells, KLF2 was shown to be negatively regulated by p53, which bound to the KLF2 promoter to induce deacetylation of the KLF2 histone H3 [34]. Tumor necrosis factor alpha (TNF-α) was shown to activate NF-Κβ p65 to complex with histone deacetylase 4 to prevent MEF2 binding to the KLF2 promoter, demonstrating a possible additional mechanism of the downregulation of KLF2 in endothelial cells in response to injury. Finally, low-density lipoprotein (LDL) cholesterol was found to stimulate the methylation of both DNA and histones on the KLF2 promoter and to contribute to the downregulation of KLF2 in response to LDL cholesterol. These mechanisms suggest there are a number of complex pathways that control the expression of KLF2 in a number of different tissue types.

2.3 Krüppel-like factor 3

Krüppel-like factor 3 (KLF3) or basic Krüppel-like factor (BKLF) is widely expressed and abundant in erythroid cells. KLF3 is believed to regulate adipogenesis, erythropoiesis, and B-cell development [35, 36]. KLF3 is able to interact with the co-repressor CtBP to repress gene transcription much like Krüppel-like factor 8 (KLF8) and Krüppel-like factor 12 (KLF12), and the N-terminal repression domain is important for this interaction in KLF3 [37–39]. KLF3 has been found to be sumoylated and that this sumoylation also affects its interaction with CtBP [37]. KLF3 has been shown to have a role in adipogenesis as forced expression of KLF3 was shown to block adipocyte differentiation [40]. Recent methylation data from endothelial cells demonstrates that KLF3 is highly methylated in flow-dependent conditions but can be reversed with 5-aza-2′-deoxycytidine treatments [41]

2.4 Krüppel-like factor 4

Krüppel-like factor 4 or gut-enriched Krüppel-like factor (GKLF) or endothelial zinc finger (EZF) protein is most similar to KLF2 and functions in the regulation of the epithelial of the gut and skin, endothelial cells, smooth muscle cells in vascular disease, and induced pluripotent stem cells (iPSC) [1, 42]. KLF4−/− mice died shortly after birth due to epithelial barrier defects in skin and gut barriers [43]. KLF4 is regulated by AP-2alpha during early and mid-embryogenesis to help regulate proliferation [44].

KLF4 became well-known after the discovery that it was one of the regulating factors along with Oct4, Sox2, and Nanog of induced pluripotent stem cells [4–7]. Oct4 was later found to regulate the expression of KLF2, while LIF/Stat3 was thought to regulate the activation of KLF4 in embryonic stem cells [45, 46]. Additional studies have suggested that posttranslational modifications increase or decrease the stability of KLF4 mRNA and these modifications control the exit from pluripotency [47]. Furthermore, these modifications mediate the ability of KLF4 to complex with other pluripotency transcription factors and bind DNA. Finally, Oct4 has been shown to contain a linker region that is important for loosening chromatin, complexing with Brg1, and allowing for KLF4 to bind during cellular reprogramming [2]. Clearly, the interactions and mechanisms of pluripotency factors in stem cells are complex and require further investigation.

KLF4 is required for normal functioning of the gut epithelial as deletion of KLF4 resulted in altered proliferation [48]. KLF4 and KLF5 are often found in the same types of tissues, bind to similar or identical DNA elements, and often exert opposing affects in different tissue types. KLF4 has been found to bind with p53 on the p21 genes in epithelial cells and in smooth muscle cells to inhibit proliferation [42, 49, 50].

In the case of smooth muscle cell proliferation, sumoylation of KLF4 causes it to fall off the p21 promoter and decreases p21 transcription following PDGF-BB treatments [51]. Sumoylation is also believed to affect binding of KLF4 to smooth muscle marker genes in TGFβ treatment [52, 53]. In smooth muscle cells in vascular disease, KLf4 has been shown to be activated by Sp1 and Oct4 binding to the KLF4 promoter [54, 55]. Separately, in macrophages KLF4 sumoylation promotes an IL-4-induced macrophage polarization to an M2 state, suggesting KLF4 plays a role in inflammation and macrophage polarization states [56]. However, in endothelial cells KLF4 is important along with KLF2 for the maintenance of endothelial cell integrity and normal endothelial barrier function [29]. KLF4 function in vascular disease could fill chapters of books investigating its many roles and functions; however, our goal is to highlight some of the mechanisms of its regulation in these processes.

Finally, KLF4 is also regulated by DNA methylation in several different types of cancers. KLF4 was found to be hypermethylated in renal cell carcinomas [57] and endometrial cancers [58]. However, a surprising discovery was KLF4 can bind to methylated regions of chromatin to mediate activation of transcription without the need for demethylation of the DNA in some types of cancer cells [59, 60]. These studies demonstrate a new role for some transcription factors as methylation readers in the transcription process.

2.5 Krüppel-like factor 5

Krüppel-like factor 5 or intestinal-enriched Krüppel-like factor (IKLF) or basic transcription element-binding protein 2 (BTEB2) is located on chromosome 13q22.1 and is important in the expression of the gut epithelia, vascular smooth muscle cells, and white adipose tissues [1, 61]. KLF5 is important in epithelial cells as it is located in the base of the crypts where cells are proliferating toward the villi. In general, KLF4 and KLF5 have been shown to compete to the same sites on DNA [62] and have also been suggested to be involved in their own regulation [42]. KLF5 has been shown to be important in gastric tumor progression and initiation and often correlate with KRAS mutations [63, 64].

KLF5 has also demonstrated to be important in the development and maintenance of the heart, aorta, and lung systems [20, 65–69]. Following angiotensin II induction, KLF5 was shown to bind to PDGF-A and activate it. KLF5 was also shown to be activated by RARα binding site in the KLF5 promoter [65, 70]. KLF5 has been shown to be regulated by acetylation. When KLF5 is associated with p300, it is acetylated and able to activate gene expression. Conversely, when SET is bound to KLF5, it prevents acetylation of KLF5 and its transcriptional activity [71]. These studies suggest that KLF5 can be regulated directly by modifications to control its transcriptional activity.

Expression of KLF5 in breast cancers was found to be correlated with a negative prognosis and decreased survival [72], while in clear cell renal cell carcinoma, hypermethylation and decreased expression of KLF5 were associated with a poorer prognosis [73]. Hypermethylation of KLF5 in acute myeloid leukemia was also associated with a poorer prognosis [74]. These studies suggest that KLF5 function in cancer is cell and perhaps even cell lineage specific. Within various cancers, KLF5 has also been demonstrated to be regulated by micro-RNAs. In gastric cancer, miR-145-5p directly targets KLF5 and promotes the differentiation of gastric cancer via KLF5 downregulation [75]. Separately, in hepatocellular carcinoma miR-214-5p acted as a tumor suppressor that could directly target and promote the downregulation of KLF5 [76]. These data demonstrate complex regulatory pathways involved in KLF5 regulation in cancer progression.

2.6 Krüppel-like factor 6

Krüppel-like factor 6 (KLF6) or zinc finger transcription factor 9 (ZF9) has been shown to be important for endothelial biology, adipogenesis, and tumor suppression in a wide variety of cancers. During embryogenesis, it is expressed in a time-sensitive manner in the kidney, cornea, gut, and yolk sac [77–80]. KLF6−/− mice are embryonic lethal due to yolk sac abnormalities [77–80]. KLF6 has been suggested to have a role in endothelial vascular remodeling following injury as it binds and activated urokinase plasminogen activator 1, endoglin, and matrix metalloproteinase 9 [81]. Interestingly, KLF6 has an alternative form of regulation because the gene produces at least four different isoforms that are able to affect DNA binding and transcription [82]. The full-length isoform of KLF6 is believed to function as a tumor suppressor and can be regulated by loss of heterozygosity, mutation, or decreased expression in different cancer types. The full-length KLf6 was found to have one deleted allele in prostate cancer, and the leftover allele was mutated 71% of the time, preventing KLF6 from functioning to activate p21 [83]. Of the isoforms of KLF6, the Krüppel-like factor 6 splice variant 1 (KLF6-SV1) was found to be oncogenic and upregulated in prostate, lung, and breast cancers and inhibits the activity of the full-length KLF6 [82]. This is the first KLF to be regulated in part by alternative splicing and suggests that directed targeting of the splice variants of KLF6 could represent a potential target for elimination therapy.

KLF6 can be regulated by methylation both to downregulate its expression and to prevent its binding to certain sites in cancer. Studies have suggested a possible role for methylation of KLF6 in hepatocellular carcinoma and in colorectal cancer [84, 85]. Separately, KLF6 can be prevented from binding on the SIRT5 promoter by the presence of DNA methylation during adipocyte differentiation [86]. KLF6 also could not bind the tissue factor pathway inhibitor-2 promoter following hypermethylation of its promoter during adipocyte formation [87].

2.7 Krüppel-like factor 7

Krüppel-like factor 7 (KLF7) or ubiquitous Krüppel-like factor (UKLF) has high expression in the brain and spinal cord and is important in the developing brain and nervous system [88]. KLF7 was identified originally in 1998, located on chromosome 2, and was believed to share a strong similarity with KLF6 [89]. Studies by Laub et al. found that KLF7 was important for upregulation of p21, repression of cyclin D1, and growth arrest in neuronal cells, thereby helping to lead to their differentiation and maturation [88]. In separate but related studies, the same laboratory found that elimination of KLF7 leads to neonatal lethality and the elimination affected areas of the olfactory, visual system, cerebral cortex, and hippocampus [90].

They also further investigated the roles of p21 and p27 and found KLF7 affected their expression in these areas during development [90]. Additional studies suggest that KLF7 regulates a number of genes in olfactory neuron development and axonal growth [91, 92]. In corneal epithelial differentiation, KLF7 was found by ChIP-sequencing to inhibit the activity of KLF4 to promote a corneal "progenitor"-like state [93].

KLF7 has also been suggested to play a role in type 2 diabetes. Studies have suggested that there are single nucleotide polymorphisms (SNPs) in the KLF7 gene that are associated with increased type 2 diabetes in Asian populations [94]. The same group further investigated the role of KLF7 and found that overexpression of KLF7 impaired the insulin production system and secretion in pancreatic beta cells while also inhibiting insulin sensitivity in the peripheral tissues [95]. KLF7 was also found to activate the TLR4/NF-kB/IL-6 pathway in adipocytes [96]. Finally, KLF7 has recently been also been found to be elevated in gastric cancers in patient samples in some populations and has been suggested to be a possible biomarker for the disease [97].

2.8 Krüppel-like factor 8

Krüppel-like factor 8 is expressed at low level in most tissue types [1]. KLF8 is a member of the same subfamily of Krüppel-like factors that includes KLF3 and KLF12 as all three KLFs recruit CtBP to repress transcription [37–39, 98]. These data also demonstrated that KLF8 needs its own DNA-binding domain to bind DNA but needs its repression domain for interaction with CtBP. KLF8 has been shown to be upregulated and activated during several types of cancers including those from ovarian, breast, and renal carcinomas [99–101]. KLF8 was also shown to activate the FHL2 gene in pancreatic cancer cells and to promote metastasis and epithelial-to-mesenchymal (EMT) transitions in pancreatic tumor cells [100, 101]. Furthermore, KLF8 was shown in gastric cancer to induce HIF-1 expression and promote epithelial-to-mesenchymal transitions in gastric cancer [102]. Finally, KLF8 methylation levels were also tested in prostate cancer cell lines but did not prove to be causally related to the progression of prostate tumors [103].

2.9 Krüppel-like factor 9

Krüppel-like factor 9 (KLF9) or basic transcription element-binding protein (BTEB) is broadly expressed, but its expression is especially high in the developing brain and thymus and in the smooth muscle of the gut and bladder [1, 104]. Interestingly, it has been demonstrated that although the mRNA for KLF9 is transcribed in many areas, the brain is the main organ where it is translated into protein [105]. The zinc fingers of the KLF9 gene are commonly now thought to be very closely related to Sp1 as they have a high sequence similarity. However, beyond their DNA-binding domains, these proteins share little sequence similarity [105]. In the brain expression, there is a thyroid hormone response element in the promoter of the KLF9 gene that accounts for its transcription and expression in the postnatal brain [105, 106]. KLF9 was also found to bind to a number of proximal promoter regions on genes important for brain function to repress transcription in hippocampal neurons [106, 107].

KLF9 expression has been noted in cancers of the mammary glands and uterus because of its ability to interact with the progesterone response elements to stimulate progesterone response elements [108, 109]. KLF9 is also required for the development of fertility in females as KLF9−/− mice were subfertile and were unable to differentiate their reproductive tissue without KLF9 [109]. KLF9−/− mice also were found to have aberrant regulation of their intestinal crypt cell proliferation and villus migration [110]. These data suggest that KLF9 also regulates the smooth muscle and the turnover of intestinal cells.

Finally, in follicular lymphoma, KLF9 was found to be hypermethylated and silenced in tumors along with a number of polycomb genes [111]. Separately, in breast cancer hypermethylation of KLF9 was correlated with a favorable cancer prognosis [112].

2.10 Krüppel-like factor 10

Krüppel-like factor 10 (KLF10) or transforming growth factor-inducible early gene 1 (TIEG1) is known as a TGFβ-inducible gene as it is rapidly induced by TGFβ treatments and then quickly returns back to basal levels [113, 114]. KLF10 is induced by multiple members of the TGFβ superfamily and then goes on to suppress Smad7 and co-activate together with Smad2. It is believed that KLF10 plays a major role in the mediation of TGFβ inhibition of cell proliferation and inflammation and induction of apoptosis [113, 115]. The rapid induction and then degradation of KLF10 are believed to be accounted for by SIAH proteasomal degradation [113]. In these studies, KLF10 was found to interact directly with SIAH which then mediates its degradation [113]. These studies suggest a protein degradation method of regulation.

KLF10 has been cited to be important in bone development and osteoporosis, adipocyte development, and heart, lung, brain, and T-cell activation [1, 116]. In adipocyte differentiation, C/EBPβ was found to bind and activate the KLF10 promoter, while KLF10 bound to the C-EBPα promoter to inhibit its activation [117]. In bone development, SNP analysis revealed that variants in the KLF10 gene were associated with bone loss in older men [118]. Conversely, studies in KLF10 null mice suggest a gender-specific role of KLF10 in the maintenance of bone density [19]. KLF10 null osteoblasts were also found to be defective in mineralization and in osteoblast support of osteoclast differentiation [119]. Finally, KLF10 null mice had impaired tendon function as adults with corresponding difficulty in tendon function [120].

In heart development, KLF10−/− mice developed cardiac hypertrophy and an increase in ventricle size and an increase in wall thickness, suggesting the importance of KLF10 to the maintenance of normal heart function [121]. KLF10 is also important in T-cell and Treg development along with TGFβ as deletion of KLF10 in T cells augmented atherosclerosis and led to impaired T-cell function [122].

KLF10 has been shown to be methylated in pancreatic cancers by DNMT1 with a correlation between methylation status and tumor grade [123]. The more the methylation and repression of the KLF10 promoter, the worse the tumor grade. These studies suggest that an important regulatory mechanism for KLF10 is also via methylation of its promoter.

2.11 Krüppel-like factor 11

Krüppel-like factor 11 (KLF11) or transforming growth factor-inducible early gene 2 (TIEG2) or FKLF is known to be expressed in the pancreas and in erythroid cells in the fetal liver. KLF11 is located in humans at chromosome 2p25 [1, 124–126]. KLF11 shares 91% homology with KLF10 in the zinc finger domain and 44% homology with the N-terminus of KLF10 [127]. These studies also demonstrated that overexpression of KLF11 inhibits cell proliferation [127] and is induced by TGFβ signaling pathways.

KLF11 contains three repression domains that are believed to be important for its repressor activities [128]. TGFβ signaling pathway induction means that KLF11 often cooperates with Smads to induce changes in transcription following TGFβ treatment. KLF11 later was found to be activated by several members of the TGFβ superfamily and not just by TGFβ treatment alone [114]. Studies have shown in neuronal cells that KLF11 regulates the transcription of the dopamine D2 receptor by complexing with p300, a histone acetylase, to promoter transcription [129]. KLF11 was also found to regulate collagen gene expression through the heterochromatin protein 1 gene-silencing pathway, as mutants defective for coupling to this epigenetic modifier lose the ability to repress COL1A2 and to prevent fibrosis in KLF11−/− mice [130]. As part of the TGFβ induction of KLF11, TGFβ induction allows KLF11 to interact with Smad3 and to repress certain promoters. In the case of pancreatic cancer, KLF11 was found to bind with Smad3 to the c-myc promoter following TFG-β treatment [131].

KLF11 is important not only for its TGFβ response but also for its associations with diabetes and obesity [132, 133]. A variant of KLF11 was found that could lead to type 2 diabetes and obesity [134]. Further studies revealed additional variants that may affect KLF11 regulation of the insulin promoter and type 2 diabetes [133]. KLF11 was also found to interact with p300 in maturity-onset diabetes of the young to induce transcriptional changes in the pancreas [135]. In converse, KLF11 can also interact with mSin3a in pancreatic cancer by repression of the Smad7 promoter [136]. Ectopic expression of KLF11 increased the sensitivity of cells to oxidative drugs [137]. Methylation of KLF11 has been suggested to be one mechanism of its downregulation in several types of cancers [138, 139].

2.12 Krüppel-like factor 12

Krüppel-like factor 12 or BETB1 was first identified in the regulation of the AP-2α gene and is located on chromosome 13q21-13q22 [140]. In the case of the AP-2α gene, KLF12 functions as a transcriptional activator and appears to relate back to KLF12's function as a marker of tumor development [141–143]. KLF12 is a marker for gastric cancer progression, and overexpression of KLF12 promotes tumor cell invasion and progression [142]. However, in lung cancer cell lines, it was shown that KLF12 was important for the regulation of anoikis and the progression through the S phase of cell cycle [141]. These data suggest that KLF12 may have multiple different roles in cancer beyond what was previously identified. KLF12 is also one of the KLF factors to interact with the mSin3a repressor complex via an alpha-helical motif in a repression domain of the transcription factor [144].

KLF12 not only plays roles in tumor progression but is also believed to play a role in the developing kidney after birth. KLF12 was shown to be expressed in the collecting ducts of the kidney after birth and could directly regulate the UT-A1 but not the ENaC promoters, two genes important for the development of the collecting ducts [145]. A recent study suggests that KLF12 might in part be regulated in cancer by the methylation of miR-205 by long noncoding RNA ELF3-antisense RNA 1. These data suggest that miR-205 and RNA ELF3-antisense RNA 1 exist in a complex regulatory loop involving KLF12 [146].

2.13 Krüppel-like factor 13

Krüppel-like factor 13 (KLF13) or BTEB3, FKLF2, or RFLAT-1 was first discovered along with Krüppel-like factor 14 (KLF14) using an expressed sequence tag database to search for additional conserved KLF DNA-binding domains [129]. KLF13−/− mice are one of the few KLF mice that are viable and fertile; however, they display abnormal blood cell development [147, 148] suggesting that KLF13 is critical for both B- and T-cell developments [148–150]. One part of this developmental process is KLF13's interaction with PPAR4 [151] to regulate CCL5. Not only is KLF13 important for blood cell development, it has also been shown to be important for the developing heart [104, 152]. To this end, KLF14 can also be linked to Holt-Oram syndrome, an inherited disorder characterized by abnormalities of the upper limbs and heart, via its interaction with the TBX5 promoter [153].

KLF13 has also recently been suggested to be a tumor suppressor in glioma cells [154]. These studies found that KLF13 was downregulated by hypomethylation across the gene to promote its silencing; however, decreases in DNMT1 expression or decreases in hypomethylation patterns of KLF13 decreased proliferation and migration of glioma cells [154]. Another example of KLF13 methylation is the methylation of the obesity-related variant of KLF13: cg07814318. The methylation of this particular SNP appears to be related to increased childhood obesity [155]. These studies suggest that methylation of promoters could be one possible mechanism of regulation of KLFs in development or disease.

Another possible mechanism of regulation of KLF13 is through the co-repressor complex mSin3a [144]. In this instance, KLF13 was found to interact with the mSin3a repressor complex via an alpha-helical motif in a repression domain [144]. Additional studies from this group suggest that multiple KLF factors (BTEB1, BTEB3, BTEB4) could also contain this alpha-helical domain in their repression regions.

2.14 Krüppel-like factor 14

Krüppel-like factor 14 was first discovered using expressed sequence tag databases to search for the presence of additional conserved KLF DNA-binding domains [129]. KLF14 has 72% similarity with the human Sp2; however, the majority of its similarity exists within its DNA-binding domain [129]. Most reports suggest that its expression is ubiquitous [1]. Interestingly, KLF14 is intron-less and exists on chromosome 7q32. KLF14 is a mono-allelic expression pattern and shown to be hypomethylated in many tissues, further suggesting a pattern of ubiquitous expression [156]. Further evidence also suggests that KLF14 could be derived from a retro-transposed copy of Krüppel-like factor 16 (KLF16) [156] and could be an example of accelerated evolution. KLF14 deletion has recently been linked with centrosome amplification, aneuploidy, and spontaneous tumorigenesis because KLF14 functions as a repressor of polo-like kinase 4 (PLK4). Without the repressive activities of KLF14 on PLK-14, PLK-14 can cause chromosomal abnormalities and promote tumorigenesis in cancer cells. The KLF14 gene has been linked to genomic variants that are highly correlative with basal cell carcinoma [157].

Genome-wide association studies not only revealed that KLF14 was linked with basal cell carcinoma, it also has revealed that KLF14 is linked with cholesterol metabolism, metabolic disease, and coronary artery disease. These studies suggest that KLF14 might function as an imprinted master regulator of metabolic function and that mutation of certain SNPs within the KLF14 gene can lead to a large-scale deregulation of metabolic gene function [158]. KLF14 was also found to regulate levels of HDL-C and hepatic ApoA-I production [159]. Guo et al. were able to find evidence that perhexiline was able to activate KLF14 and to reduce lesions in ApoE−/− atherosclerotic mice [159]. Separate but related studies suggest that this activity is related to the phosphorylation of KLF14 by both p38 MAPK and ERK kinase [160]. However, KLF14 was found to be decreased in endothelial cells in atherosclerosis, and overexpression of KLF14 actually inhibited NF-KB signaling by suppressing p65 [161]. KLF14 has also been shown to interact with p300 to promote sphingosine kinase activation and to enhance sphingosine production [162].

These data suggest a complicated pattern of expression for a ubiquitous transcription factor that could produce paradoxical effects in inflammatory disease such as cardiovascular disease or cancer. Interestingly, there still appears to be less known about how KLF14 itself is regulated.

2.15 Krüppel-like factor 15

KLF15 or kidney-enriched Krüppel-like factor (KKLF) demonstrates low levels of cardiac-specific expression during development but then exhibits adult expression in the kidney, liver, pancreas, heart, skeletal muscle, lung, and ovary. KLF15 was originally thought to be important for the regulation of different cell types in the kidney and repressed genes such as CLC-K1 and CLC-K2 [163]. However, its regulatory effects can be seen in the heart, skeletal muscle, gluconeogenesis, and circadian rhythms. In terms of the heart, KLF15 was demonstrated to be an inhibitor of cardiac fibrosis by repression of connective tissue growth factor (CTGF) [164]. In this mechanism, KLF15 inhibits the recruitment of the co-activator P-CAF but does not prevent SMAD3 from binding to the promoter [164]. Additional studies by the same group demonstrated that KLF15 was a negative regulator of cardiac hypertrophy via inhibition of GATA4 and MEF2 functions [165]. Recent studies further suggest that KLF15 was identified as a putative upstream regulator of metabolic gene expression in the heart via RNA-Seq and methylation sequencing and that KLF15 was itself regulated by EZH2 in a SET domain-dependent manner [166]. KLF15 was demonstrated to be silenced via methylation in ischemic cardiomyopathy which in turn leads to the silencing of many cardiac-specific genes.

KLF15 has been shown to also be important for metabolism [167]. In terms of the skeletal muscle, overnight fasting and endurance exercise induce KLF15 expression, while knockout of KLF15 induces abnormal energy flux, excessive muscle fatigue, and impaired endurance capacity [168]. KLF15 was later shown to complex in the liver with liver X receptor (LXR) to inhibit SREBF1 during fasting by recruiting the co-repressor RIP140 [169]. Finally, KLF15 is also important for nitrogen homeostasis and the maintenance of circadian rhythm as KLF15 knockout mice had no amino acid rhythm and no rhythm of the production of urea from ammonia [170]. These studies suggest the importance of KLF15 and suggest that investigations into how it is regulated by chromatin readers and writers will become important to these metabolic diseases.

2.16 Krüppel-like factor 16

Krüppel-like factor 16 or dopamine receptor regulating factor (DRRF) was first discovered in its regulation of the dopamine receptors in the developing brain and eye [171]. It is now known that KLF16 is expressed not only in the developing brain but also in the thymus, intestine, kidney, liver, heart, and bladder. KLF16 has recently been shown to not only regulate the dopamine receptor but also to regulate the ephrin receptor A5 (EphA5), but this regulation was methylation specific as methylation of the EphA5 promoter prevented KLF16 from binding [171]. These data suggest that one possible epigenetic mechanism regulating KLF16 is methylation of regions near its binding site.

KLF16 was found by Daftary et al. to bind to all three types of KLF binding site, the GC, CA, and BTE boxes using electromobility shift assays but prefers binding to the BTE box in cells and to mediate its effects via mSin3a, a transcriptional co-repressor complex but suggests that this function is both promoter and cell context dependent [172]. To further study this interaction, site-directed mutagenesis was performed of all of the serine, threonine, and tyrosine residues believed to be possible targets for kinase phosphorylation signaling and found that mutation of tyrosine-10 altered the ability of KLF16 to interact with mSin3a [172]. Finally, KLF16 was also found to be regulated by nuclear localization and to be excluded from heterochromatin within the nucleus [172]. These studies suggest complex posttranslational regulatory mechanisms for KLF16 function in a cell- and promoter-dependent manner.

2.17 Krüppel-like factor 17

Krüppel-like factor 17 (KLF17) was first discovered in mouse as zinc finger protein 393 (ZFP393) or ZNF393 where it was shown to be expressed in the testis and ovaries, and the gene spans 8 kb in the distal portion of chromosome 4 in the mouse [173]. In humans KLF17 maps to chromosome 1p34.1. When it was discovered back in 2002, it was believed to be the first C2H2 germ cell-specific zinc finger protein. Identification of KLF17 in the human revealed that KLF17 was expressed not only in the testis but also in the brain and bone, albeit at relatively low amounts [174]. KLF17 also contains low sequence similarity between the human and mouse orthologues; however, a detailed transcriptional binding analysis by van Vliet et al. was able to demonstrate that KLF17 was a Krüppel-like transcription factor rather than being more closely linked to the specificity protein factor family (Sp family) [173].

KLF17 is hypothesized to be a tumor suppressor in multiple types of cancers, and a decrease in its expression has become correlated with a poor cancer prognosis [175]. KLF17 was demonstrated to be a tumor suppressor gene in metastatic breast cancer lines whose downregulation promotes the epithelial-to-mesenchymal transition in cancer cells [176]. These studies also suggested that KLF17 is a direct negative regulator of inhibitor of DNA binding 1 (ID1). Sadly, they do not offer a direct mechanism for the downregulation of KLF17 during breast cancer metastasis, but they do provide compelling data to suggest that KLF17 might have multiple functions in the male and female sex organs and that suppression of this factor could lead to increased tumorigenic potential [176].

Further evidence in non-small-cell lung cancer also suggests that KLF17 could function as a tumor suppressor [177]. These studies suggested that p53 recruits p300 to the KLF17 promoter to acetylate and turn on transcription [177]. In addition, p53 also physically interacts with KLF17 and promotes binding of KLF17 to certain gene promoters and promotes transcription of p53, p21, and pRB [177]. These data suggest an intricate cross-talk between KLF17 and p53 in tumorigenesis. Another way KLF17 is believed to inhibit cancer progression is through inhibition of proliferation via repression of UPAI-1 [178], which Cai et al. proposed inhibited the invasive properties of small-cell lung cancer cells. KLF17 was also suggested to be a tumor suppressor through a TGFB-/SMAD-dependent mechanism where KLF17 physically interacts with SMAD3 to target genes to prevent metastases [179]. MiR-9, a micro-RNA important for tumor invasion and metastasis, has been shown to inhibit the activation of KLF17 by directly binding to the 3′-untranslated region (3′-UTR) [175]. These pathways suggest that KLF17 can be regulated both by direct promoter activation and by posttranscriptional modifications such as RNA degradation by micro-RNAs.

In converse, in endometrial cancer KLF17 was found to be an inducer of epithelial-to-mesenchymal transition and resulted in activation of TWIST1 [180]. This finding demonstrated that KLF17 bound directly to the TWIST promoter to activate its transcription [180]. KLF17 was also shown to bind directly to estrogen receptor alpha (ERα) to prevent it from being able to bind directly to chromatin [181]. ERα then also contributed to the suppression of KLF17 using the co-repressor histone deacetylase 1 (HDAC1) to promote KLF17 deacetylation and chromatin condensation [181].

2.18 Krüppel-like factor 18

Krüppel-like factor 18 (KLF18) was identified in 2013 from sequence similarity searches and gene synteny analyses and was shown at that time to be highly related to KLF17 [182]. Like KLF17, it is believed to be expressed in the developing testis and restricted to that area. Little data currently exists examining its function; however, a detailed analysis of its structure and phylogenic tree in placental mammals has been investigated in detail by Pei et al. [182]. This group also suggested that KLF18 might be a pseudogene of KLF17 since its expression pattern is restricted and it is similar in sequence to KLF17. Despite this hypothesis, three genes in mouse and rat were identified that closely resemble KLF18: Zfp352, Zfp352-like, and Zfp353 [182]. The promoter and/or details into the transcriptional activation of this KLF are currently unknown. A more detailed analysis of the functions and regulations of KLF18 would provide more insight into this transcription factor's function.

3. Concluding remarks

Over the past 20 years since the discovery of the first KLF transcription factor, there continues to be a growing body of evidence to suggest that KLFs are important to tumor progression, cardiovascular disease, metabolism, and even circadian rhythm [1]. While much of the work has focused on the functions of these factors and their roles in various disease processes, there still remains additional needed work to explain how the various KLFs become activated and/or repressed during diseased states. There is a growing body of evidence, which we have attempted to discuss in some detail in this chapter, in the more extensively studied KLFs such as KLF4, KLF5, and KLF2 that suggest that the KLFs are regulated extensively by posttranslational modifications such as phosphorylation, acetylation, ubiquitination, and sumoylation. These modifications appear to be critical for co-factor recruitment and determination of whether KLFs interact with either activators or repressors of transcription. It has been interesting to see the wealth of information that has developed over the past 20 years investigating the roles of these various factors in various diseases; however, relatively speaking, we still know little about how these factors are activated and/or repressed transcriptionally during diseased states.

Since the onset of the era of big data, more of the KLF field has come to focus on the roles of pathway analysis following genetic ablation of a KLF in a cell-specific manner. These studies have yielded enormous amounts of data that offer valuable insight into the overlap between various KLF factors in diseases [183]. It will be of interest in the future to see how the integration of single-cell genomics will come into play with various different roles of the same KLF in various cell types in diseased states [184]. For example, the integration of single-cell RNA-Seq [184] with Assay for Transposase-Accessible Chromatin using sequencing (ATAC-Seq) [185, 186] in cells where a single KLF bear separate functions could offer deeper insight of the role of the niche environment on KLF function and/or on the roles of KLFs in downstream activations of different types of pathways during disease. Cardiovascular diseases have recently begun to investigate single-cell sequencing with other factors, such as Tcf21, and were able to use these innovative studies to investigate the role of this factor in smooth muscle cell to fibroblast transitions during atherosclerosis [184]. It will be exciting to see how KLF biology will use this technology to further investigate how these transcription factors regulate disease.

Not only will the integration of single-cell studies with KLF function give us greater insight into KLF function in development and disease, but the study of the role of RNA posttranscriptional modifications will most likely play an emerging role in the KLF field in the near future [184]. Since the sequencing of the human genome and the growing realization of the stronger role of RNA in transcriptional and translational control, there has been a re-emergence of interest in the field of RNA posttranscriptional modifications [187]. There are over 100 different types of RNA modifications of which the N^6-methyladenosine (m^6A) modification is the most common [187]. Interestingly, m^6A has recently been shown to be concentrated in the 3′-UTR of many messenger RNAs and that micro-RNAs are capable of mediating this modification via a sequence pairing mechanism to help promote stem cell pluripotency [187–192]. This new role for RNA modification and stem cell maintenance has immense implications for KLFs involved in induced pluripotent stem maintenance like KLF4. Therefore, it will be of interest to determine whether RNA modifications affect other disease processes by similar sequence pairing mechanisms.

In conclusion, the KLF field has offered many insights to different disease processes since the discovery of the first KLF over the past 20 years. New insights into the regulation of these factors will hopefully grant novel methods to directly and properly target these factors to inhibit diseased states that currently have no medical treatment therapy. Perhaps the newly emerging CRISP technology will be able to directly target KLFs in a cell-specific manner as many KLFs have opposing functions in many different cell types. In any case, this transcription factor family has offered much excitement since its discovery and hopefully will offer new insights as the field studies these factors in more depth in the future.

Acknowledgements

This work was supported by the AHA Scientist Development Grant 14SDG18730000 (MS). The content is solely the responsibility of the authors and does not necessarily represent the views of the AHA.

Conflict of interest

The authors declare no conflict of interest.

Notes/thanks/other declarations

We thank Anthony Herring and Cindy Dodson for their knowledge and technical expertise.

Nomenclatures

DNMT1	DNA methyltransferase 1
EphA5	Ephrin receptor A5
EMT	Epithelial-to-mesenchymal transition
ER	Estrogen receptor
HDAC	Histone deacetylase
ID1	Inhibitor of DNA binding 1

IL-4	Interleukin-4
IL-6	Interleukin-6
NF-KB	Nuclear Factor kappa-light-chain-enhancer of activated B cells
KLF	Krüppel-like factor
M6A	N^6-methyladenosine
mSin3A	Co-repressor complex used for repression
p300	Histone acetylase
P53	TP53 or tumor protein
P50	Subunit of NF-KB
P65	Subunit of NF-KB signaling
P21	p21CIP1, cyclin-dependent protein inhibitor
PDGF-BB	Platelet-derived growth factor BB
pRB	Phosphorylated RB
SMC	Smooth muscle cells
Smad	Proteins transduce signals from transforming growth factor beta
SM-actin	Smooth muscle alpha actin
Sp	Specificity proteins
TFG-β	Transforming growth factor beta
TNF-α	Tumor necrosis factor alpha
TWIST	TWIST1-protein
ZF-TF	Zinc finger transcription factor
ZFP	Zinc finger protein

Author details

Morgan Salmon
Department of Surgery, University of Virginia School of Medicine, Charlottesville, VA, USA

*Address all correspondence to: msa5m@virginia.edu

References

[1] McConnell BB, Yang VW. Mammalian Krüppel-like factors in health and diseases. Physiological Reviews. 2010;**90**:1337-1381

[2] Chen K, Long Q, Xing G, Wang T, Wu Y, Li L, et al. Heterochromatin loosening by the Oct4 linker region facilitates Klf4 binding and iPSC reprogramming. The EMBO Journal. 2020;**39**:e99165

[3] Homma K, Sone M, Taura D, Yamahara K, Suzuki Y, Takahashi K, et al. Sirt1 plays an important role in mediating greater functionality of human ES/iPS-derived vascular endothelial cells. Atherosclerosis. 2010;**212**:42-47

[4] Nakagawa M, Koyanagi M, Tanabe K, Takahashi K, Ichisaka T, Aoi T, et al. Generation of induced pluripotent stem cells without Myc from mouse and human fibroblasts. Nature Biotechnology. 2008;**26**:101-106

[5] Narazaki G, Uosaki H, Teranishi M, Okita K, Kim B, Matsuoka S, et al. Directed and systematic differentiation of cardiovascular cells from mouse induced pluripotent stem cells. Circulation. 2008;**118**:498-506

[6] Takahashi K, Tanabe K, Ohnuki M, Narita M, Ichisaka T, Tomoda K, et al. Induction of pluripotent stem cells from adult human fibroblasts by defined factors. Cell. 2007;**131**:861-872

[7] Takahashi K, Yamanaka S. Induction of pluripotent stem cells from mouse embryonic and adult fibroblast cultures by defined factors. Cell. 2006;**126**:663-676

[8] Nuez B, Michalovich D, Bygrave A, Ploemacher R, Grosveld F. Defective haematopoiesis in fetal liver resulting from inactivation of the EKLF gene. Nature. 1995;**375**:316-318

[9] Asano H, Stamatoyannopoulos G. Activation of beta-globin promoter by erythroid Krüppel-like factor. Molecular and Cellular Biology. 1998;**18**:102-109

[10] Perkins AC, Peterson KR, Stamatoyannopoulos G, Witkowska HE, Orkin SH. Fetal expression of a human Agamma globin transgene rescues globin chain imbalance but not hemolysis in EKLF null mouse embryos. Blood. 2000;**95**:1827-1833

[11] Perkins AC, Sharpe AH, Orkin SH. Lethal beta-thalassaemia in mice lacking the erythroid CACCC-transcription factor EKLF. Nature. 1995;**375**:318-322

[12] Tallack MR, Perkins AC. Three fingers on the switch: Krüppel-like factor 1 regulation of γ-globin to β-globin gene switching. Current Opinion in Hematology. 2013;**20**:193-200

[13] Cantú I, van de Werken HJG, Gillemans N, Stadhouders R, Heshusius S, Maas A, et al. The mouse KLF1 Nan variant impairs nuclear condensation and erythroid maturation. PLoS One. 2019;**14**:e0208659-e

[14] Fluhr S, Krombholz CF, Meier A, Epting T, Mücke O, Plass C, et al. Epigenetic dysregulation of the erythropoietic transcription factor KLF1 and the β-like globin locus in juvenile myelomonocytic leukemia. Epigenetics. 2017;**12**:715-723

[15] Frontelo P, Manwani D, Galdass M, Karsunky H, Lohmann F, Gallagher PG, et al. Novel role for EKLF in megakaryocyte lineage commitment. Blood. 2007;**110**:3871-3880

[16] Wani MA, Conkright MD, Jeffries S, Hughes MJ, Lingrel JB. cDNA isolation, genomic structure, regulation, and chromosomal localization of human

lung Krüppel-like factor. Genomics. 1999;**60**:78-86

[17] Basu P, Morris PE, Haar JL, Wani MA, Lingrel JB, Gaensler KML, et al. KLF2 is essential for primitive erythropoiesis and regulates the human and murine embryonic beta-like globin genes in vivo. Blood. 2005;**106**:2566-2571

[18] Carlson CM, Endrizzi BT, Wu J, Ding X, Weinreich MA, Walsh ER, et al. Krüppel-like factor 2 regulates thymocyte and T-cell migration. Nature. 2006;**442**:299-302

[19] Hawse JR, Iwaniec UT, Bensamoun SF, Monroe DG, Peters KD, Ilharreborde B, et al. TIEG-null mice display an osteopenic gender-specific phenotype. Bone. 2008;**42**:1025-1031

[20] Lin S-CJ, Wani MA, Whitsett JA, Wells JM. Klf5 regulates lineage formation in the pre-implantation mouse embryo. Development (Cambridge, England). 2010;**137**: 3953-3963

[21] Wani M, Means R, Lingrel J. Loss of LKLF function results in embryonic lethality in mice. Transgenic Research. 1998;7(4):229-238

[22] Wani MA, Wert SE, Lingrel JB. Lung Krüppel-like factor, a zinc finger transcription factor, is essential for normal lung development. The Journal of Biological Chemistry. 1999;**274**:21180-21185

[23] Jiang W, Xu X, Deng S, Luo J, Xu H, Wang C, et al. Methylation of krüppel-like factor 2 (KLF2) associates with its expression and non-small cell lung cancer progression. American Journal of Translational Research. 2017;**9**:2024-2037

[24] Kuo CT, Veselits ML, Leiden JM. LKLF: A transcriptional regulator of single-positive T cell quiescence and survival. Science (New York, NY). 1997;**277**:1986-1990

[25] Ohnesorge N, Viemann D, Schmidt N, Czymai T, Spiering D, Schmolke M, et al. Erk5 activation elicits a vasoprotective endothelial phenotype via induction of Krüppel-like factor 4 (KLF4). The Journal of Biological Chemistry. 2010;**285**:26199-26210

[26] Sohn SJ, Li D, Lee LK, Winoto A. Transcriptional regulation of tissue-specific genes by the ERK5 mitogen-activated protein kinase. Molecular and Cellular Biology. 2005;**25**:8553-8566

[27] Chiplunkar AR, Lung TK, Alhashem Y, Koppenhaver BA, Salloum FN, Kukreja RC, et al. Krüppel-like factor 2 is required for normal mouse cardiac development. PLoS One. 2013;**8**:e54891

[28] Lee JS, Yu Q, Shin JT, Sebzda E, Bertozzi C, Chen M, et al. Klf2 is an essential regulator of vascular hemodynamic forces in vivo. Developmental Cell. 2006;**11**:845-857

[29] Sangwung P, Zhou G, Nayak L, Chan ER, Kumar S, Kang DW, et al. KLF2 and KLF4 control endothelial identity and vascular integrity. JCI Insight. 2017;**2**:e91700

[30] Huddleson JP, Srinivasan S, Ahmad N, Lingrel JB. Fluid shear stress induces endothelial KLF2 gene expression through a defined promoter region. Biological Chemistry. 2004;**385**:723-729

[31] Huddleson JP, Ahmad N, Srinivasan S, Lingrel JB. Induction of KLF2 by fluid shear stress requires a novel promoter element activated by a phosphatidylinositol 3-kinase-dependent chromatin-remodeling pathway. The Journal of Biological Chemistry. 2005;**280**:23371-23379

[32] Ahmad N, Lingrel JB. Krüppel-like factor 2 transcriptional

regulation involves heterogeneous nuclear ribonucleoproteins and acetyltransferases. Biochemistry. 2005;**44**:6276-6285

[33] Huddleson JP, Ahmad N, Lingrel JB. Up-regulation of the KLF2 transcription factor by fluid shear stress requires nucleolin. The Journal of Biological Chemistry. 2006;**281**:15121-15128

[34] Kumar A, Kim C-S, Hoffman TA, Naqvi A, Dericco J, Jung S-B, et al. p53 impairs endothelial function by transcriptionally repressing Krüppel-like factor 2. Arteriosclerosis, Thrombosis, and Vascular Biology. 2011;**31**:133-141

[35] Crossley M, Whitelaw E, Perkins A, Williams G, Fujiwara Y, Orkin SH. Isolation and characterization of the cDNA encoding BKLF/TEF-2, a major CACCC-box-binding protein in erythroid cells and selected other cells. Molecular and Cellular Biology. 1996;**16**:1695-1705

[36] Vu T, Gatto D, Turner V, Funnell A, Mak KS, Norton L, et al. Impaired B Cell Development in the Absence of Krüppel-like factor 3. Journal of Immunology (Baltimore, MD: 1950). 2011;**187**:5032-5042

[37] Pearson RCM, Funnell APW, Crossley M. The mammalian zinc finger transcription factor Krüppel-like factor 3 (KLF3/BKLF). IUBMB Life. 2011;**63**:86-93

[38] Turner J, Crossley M. Cloning and characterization of mCtBP2, a co-repressor that associates with basic Krüppel-like factor and other mammalian transcriptional regulators. The EMBO Journal. 1998;**17**:5129-5140

[39] Turner J, Nicholas H, Bishop D, Matthews JM, Crossley M. The LIM protein FHL3 binds basic Krüppel-like factor/Krüppel-like factor 3 and its co-repressor C-terminal-binding protein 2. The Journal of Biological Chemistry. 2003;**278**:12786-12795

[40] Sue N, Jack BHA, Eaton SA, Pearson RCM, Funnell APW, Turner J, et al. Targeted disruption of the basic Krüppel-like factor gene (Klf3) reveals a role in adipogenesis. Molecular and Cellular Biology. 2008;**28**:3967-3978

[41] Dunn J, Thabet S, Jo H. Flow-dependent epigenetic DNA methylation in endothelial gene expression and atherosclerosis. Arteriosclerosis, Thrombosis, and Vascular Biology. 2015;**35**:1562-1569

[42] McConnell BB, Ghaleb AM, Nandan MO, Yang VW. The diverse functions of Krüppel-like factors 4 and 5 in epithelial biology and pathobiology. BioEssays: News and Reviews in Molecular, Cellular and Developmental Biology. 2007;**29**:549-557

[43] Segre J, Bauer C, Fuchs E. KLF4 is a transcription factor required for establishing the barrier function of the skin. Nature Genetics. 1999;**22**:356-60

[44] Ehlermann J, Pfisterer P, Schorle H. Dynamic expression of Krüppel-like factor 4 (Klf4), a target of transcription factor AP-2alpha during murine mid-embryogenesis. The Anatomical Record Part A, Discoveries in Molecular, Cellular, and Evolutionary Biology. 2003;**273**:677-680

[45] Hall J, Guo G, Wray J, Eyres I, Nichols J, Grotewold L, et al. Oct4 and LIF/Stat3 additively induce Krüppel factors to sustain embryonic stem cell self-renewal. Cell Stem Cell. 2009;**5**:597-609

[46] Jiang J, Chan Y-S, Loh Y-H, Cai J, Tong G-Q, Lim C-A, et al. A core Klf circuitry regulates self-renewal of embryonic stem cells. Nature Cell Biology. 2008;**10**:353-360

[47] Dhaliwal NK, Abatti LE, Mitchell JA. KLF4 protein stability

regulated by interaction with pluripotency transcription factors overrides transcriptional control. Genes and Development. 2019:**33**(15-16):1069-1082

[48] Katz JP, Perreault N, Goldstein BG, Actman L, McNally SR, Silberg DG, et al. Loss of Klf4 in mice causes altered proliferation and differentiation and precancerous changes in the adult stomach. Gastroenterology. 2005;**128**:935-945

[49] Ghaleb AM, Nandan MO, Chanchevalap S, Dalton WB, Hisamuddin IM, Yang VW. Krüppel-like factors 4 and 5: the yin and yang regulators of cellular proliferation. Cell Research. 2005;**15**:92-96

[50] Yoshida T, Kaestner KH, Owens GK. Conditional deletion of Krüppel-like factor 4 delays downregulation of smooth muscle cell differentiation markers but accelerates neointimal formation following vascular injury. Circulation Research. 2008;**102**:1548-1557

[51] Nie C-J, Li YH, Zhang X-H, Wang Z-P, Jiang W, Zhang Y, et al. SUMOylation of KLF4 acts as a switch in transcriptional programs that control VSMC proliferation. Experimental Cell Research. 2016;**342**:20-31

[52] Liu Y, Sinha S, Owens G. A transforming growth factor-b control element required for SM a-actin expression in vivo also partially mediates GKLF-dependent transcriptional repression. The Journal of Biological Chemistry. 2003;**278**:48004-48011

[53] Liu Y, Sinha S, McDonald OG, Shang Y, Hoofnagle MH, Owens GK. Kruppel-like factor 4 abrogates myocardin-induced activation of smooth muscle gene expression. The Journal of Biological Chemistry. 2005;**280**(10):9719-9727

[54] Cherepanova OA, Gomez D, Shankman LS, Swiatlowska P, Williams J, Sarmento OF, et al. Activation of the pluripotency factor OCT4 in smooth muscle cells is atheroprotective. Nature Medicine. 2016;**22**:657-665

[55] Deaton RA, Gan Q, Owens GK. Sp1-dependent activation of KLF4 is required for PDGF-BB-induced phenotypic modulation of smooth muscle. American Journal of Physiology—Heart and Circulatory Physiology. 2009;**296**:H1027-H1H37

[56] Wang K, Zhou W, Cai Q, Cheng J, Cai R, Xing R. SUMOylation of KLF4 promotes IL-4 induced macrophage M2 polarization. Cell Cycle (Georgetown, Tex). 2017;**16**:374-81

[57] Li H, Wang J, Xiao W, Xia D, Lang B, Yu G, et al. Epigenetic alterations of Krüppel-like factor 4 and its tumor suppressor function in renal cell carcinoma. Carcinogenesis. 2013;**34**:2262-2270

[58] Danková Z, Braný D, Dvorská D, Ňachajová M, Fiolka R, Grendár M, et al. Methylation status of KLF4 and HS3ST2 genes as predictors of endometrial cancer and hyperplastic endometrial lesions. International Journal of Molecular Medicine. 2018;**42**:3318-3328

[59] Oyinlade O, Wei S, Kammers K, Liu S, Wang S, Ma D, et al. Analysis of KLF4 regulated genes in cancer cells reveals a role of DNA methylation in promoter- enhancer interactions. Epigenetics. 2018;**13**:751-768

[60] Wan J, Su Y, Song Q, Tung B, Oyinlade O, Liu S, et al. Methylated cis-regulatory elements mediate KLF4-dependent gene transactivation and cell migration. eLife. 2017;**6**:e20068

[61] Conkright MD, Wani MA, Anderson KP, Lingrel JB. A gene

encoding an intestinal-enriched member of the Krüppel-like factor family expressed in intestinal epithelial cells. Nucleic Acids Research. 1999;**27**:1263-1270

[62] Dang DT, Zhao W, Mahatan CS, Geiman DE, Yang VW. Opposing effects of Krüppel-like factor 4 and Krüppel-like factor 5 on the promoter of the Krüppel-like factor 4 gene. Nucleic Acids Research. 2002;**30**:2736-2741

[63] Nandan MO, Ghaleb AM, McConnell BB, Patel NV, Robine S, Yang VW. Krüppel-like factor 5 is a crucial mediator of intestinal tumorigenesis in mice harboring combined ApcMin and KRASV12 mutations. Molecular Cancer. 2010;**9**:63

[64] Nandan MO, McConnell BB, Ghaleb AM, Bialkowska AB, Sheng H, Shao J, et al. Krüppel-like factor 5 mediates cellular transformation during oncogenic KRAS-induced intestinal tumorigenesis. Gastroenterology. 2008;**134**:120-130

[65] Nagai R, Shindo T, Manabe I, Suzuki T, Kurabayashi M. KLF5/BTEB2, a Krüppel-like zinc-finger type transcription factor, mediates both smooth muscle cell activation and cardiac hypertrophy. Advances in Experimental Medicine and Biology. 2003;**538**:57-66

[66] Nagai R, Suzuki T, Aizawa A, Shindo T, Manabe I. Significance of the transcription factor KLF5 in cardiovascular remodeling. Journal of Thrombosis and Haemostasis. Aug 2005;**3**(8):1569-1576

[67] Oishi Y, Manabe I, Tobe K, Tsushima K, Shindo T, Fujiu K, et al. Krüppel-like transcription factor KLF5 is a key regulator of adipocyte differentiation. Cell Metabolism. 2005;**1**:27-39

[68] Suzuki T, Sawaki D, Aizawa K, Munemasa Y, Matsumura T, Ishida J, et al. Krüppel-like factor 5 shows proliferation-specific roles in vascular remodeling, direct stimulation of cell growth, and inhibition of apoptosis. The Journal of Biological Chemistry. 2009;**284**:9549-9557

[69] Wan H, Luo F, Wert SE, Zhang L, Xu Y, Ikegami M, et al. Krüppel-like factor 5 is required for perinatal lung morphogenesis and function. Development (Cambridge, England). 2008;**135**:2563-72

[70] Shindo T, Manabe I, Fukushima Y, Tobe K, Aizawa K, Miyamoto S, et al. Krüppel-like zinc-finger transcription factor KLF5/BTEB2 is a target for angiotensin II signaling and an essential regulator of cardiovascular remodeling. Nature Medicine. 2002;**8**:856-863

[71] Miyamoto S, Suzuki T, Muto S, Aizawa K, Kimura A, Mizuno Y, et al. Positive and negative regulation of the cardiovascular transcription factor KLF5 by p300 and the oncogenic regulator SET through interaction and acetylation on the DNA-binding domain. Molecular and Cellular Biology. 2003;**23**:8528-8541

[72] Tong D, Czerwenka K, Heinze G, Ryffel M, Schuster E, Witt A, et al. Expression of KLF5 is a prognostic factor for disease-free survival and overall survival in patients with breast cancer. Clinical Cancer Research: An Official Journal of the American Association for Cancer Research. 2006;**12**:2442-2448

[73] Fu R-J, He W, Wang X-B, Li L, Zhao H-B, Liu X-Y, et al. DNMT1-maintained hypermethylation of Krüppel-like factor 5 involves in the progression of clear cell renal cell carcinoma. Cell Death & Disease. 2017;**8**:e2952-e

[74] Diakiw SM, Perugini M, Kok CH, Engler GA, Cummings N, To LB, et al. Methylation of KLF5 contributes to reduced expression in acute myeloid

leukaemia and is associated with poor overall survival. British Journal of Haematology. 2013;**161**:884-888

[75] Zhou T, Chen S, Mao X. miR-145-5p affects the differentiation of gastric cancer by targeting KLF5 directly. Journal of Cellular Physiology. 2019;**234**:7634-7644

[76] Pang J, Li Z, Wang G, Li N, Gao Y, Wang S. miR-214-5p targets KLF5 and suppresses proliferation of human hepatocellular carcinoma cells. Journal of Cellular Biochemistry. 2018. DOI: 10.1002/jcb.27498

[77] Fischer EA, Verpont MC, Garrett-Sinha LA, Ronco PM, Rossert JA. Klf6 is a zinc finger protein expressed in a cell-specific manner during kidney development. Journal of the American Society of Nephrology. 2001;**12**(4):726-735

[78] Laub F, Aldabe R, Ramirez F, Friedman S. Embryonic expression of Krüppel-like factor 6 in neural and non-neural tissues. Mechanisms of Development. 2001;**106**:167-170

[79] Matsumoto N, Kubo A, Liu H, Akita K, Laub F, Ramirez F, et al. Developmental regulation of yolk sac hematopoiesis by Krüppel-like factor 6. Blood. 2006;**107**:1357-1365

[80] Nakamura H, dr Chiambaretta F, Sugar J, Sapin V, Yue BYJT. Developmentally regulated expression of KLF6 in the mouse cornea and lens. Investigative Ophthalmology & Visual Science. 2004;**45**:4327-32

[81] Atkins GB, Jain MK. Role of Krüppel-like transcription factors in endothelial biology. Circulation Research. 2007;**100**:1686-1695

[82] DiFeo A, Martignetti JA, Narla G. The role of KLF6 and its splice variants in cancer therapy. Drug Resistance Updates: Reviews and Commentaries in Antimicrobial and Anticancer Chemotherapy. 2009;**12**:1-7

[83] Narla G, Heath KE, Reeves HL, Li D, Giono LE, Kimmelman AC, et al. KLF6, a candidate tumor suppressor gene mutated in prostate cancer. Science (New York, NY). 2001;**294**:2563-6

[84] Song J, Kim CJ, Cho YG, Kim SY, Nam SW, Lee SH, et al. Genetic and epigenetic alterations of the KLF6 gene in hepatocellular carcinoma. Journal of Gastroenterology and Hepatology. 2006;**21**:1286-1289

[85] Babaei K, Khaksar R, Zeinali T, Hemmati H, Bandegi A, Samidoust P, et al. Epigenetic profiling of MUTYH, KLF6, WNT1 and KLF4 genes in carcinogenesis and tumorigenesis of colorectal cancer. Biomedicine. 2019;**9**:22

[86] Hong J, Wang X, Mei C, Zan L. Competitive regulation by transcription factors and DNA methylation in the bovine SIRT5 promoter: Roles of E2F4 and KLF6. Gene. 2019;**684**:39-46

[87] Guo H, Lin Y, Zhang H, Liu J, Zhang N, Li Y, et al. Tissue factor pathway inhibitor-2 was repressed by CpG hypermethylation through inhibition of KLF6 binding in highly invasive breast cancer cells. BMC Molecular Biology. 2007;**8**:110

[88] Laub F, Aldabe R, Friedrich V Jr, Ohnishi S, Yoshida T, Ramirez F. Developmental expression of mouse Krüppel-like transcription factor KLF7 suggests a potential role in neurogenesis. Developmental Biology. 2001;**233**:305-318

[89] Matsumoto N, Laub F, Aldabe R, Zhang W, Ramirez F, Yoshida T, et al. Cloning the cDNA for a new human zinc finger protein defines a group of closely related Krüppel-like transcription factors. The Journal of Biological Chemistry. 1998;**273**:28229-28237

[90] Laub F, Lei L, Sumiyoshi H, Kajimura D, Dragomir C, Smaldone S, et al. Transcription factor KLF7 is important for neuronal morphogenesis in selected regions of the nervous system. Molecular and Cellular Biology. 2005;**25**:5699-5711

[91] Kajimura D, Dragomir C, Ramirez F, Laub F. Identification of genes regulated by transcription factor KLF7 in differentiating olfactory sensory neurons. Gene. 2007;**388**:34-42

[92] Caiazzo M, Colucci-D'Amato L, Volpicelli F, Speranza L, Petrone C, Pastore L, et al. Krüppel-like factor 7 is required for olfactory bulb dopaminergic neuron development. Experimental Cell Research. 2011;**317**:464-473

[93] Klein RH, Hu W, Kashgari G, Lin Z, Nguyen T, Doan M, et al. Characterization of enhancers and the role of the transcription factor KLF7 in regulating corneal epithelial differentiation. The Journal of Biological Chemistry. 2017;**292**:18937-18950

[94] Kanazawa A, Kawamura Y, Sekine A, Iida A, Tsunoda T, Kashiwagi A, et al. Single nucleotide polymorphisms in the gene encoding Krüppel-like factor 7 are associated with type 2 diabetes. Diabetologia. 2005;**48**:1315-1322

[95] Kawamura Y, Tanaka Y, Kawamori R, Maeda S. Overexpression of Krüppel-like factor 7 regulates adipocytokine gene expressions in human adipocytes and inhibits glucose-induced insulin secretion in pancreatic beta-cell line. Molecular endocrinology (Baltimore, MD). 2006;**20**:844-56

[96] Zhang M, Wang C, Wu J, Ha X, Deng Y, Zhang X, et al. The effect and mechanism of KLF7 in the TLR4/NF-κB/IL-6 inflammatory signal pathway of adipocytes. Mediators of Inflammation. 2018;**2018**:1756494

[97] Jiang Z, Yu T, Fan Z, Yang H, Lin X. Krüppel-like factor 7 is a marker of aggressive gastric cancer and poor prognosis. Cellular Physiology and Biochemistry. 2017;**43**:1090-1099

[98] van Vliet J, Turner J, Crossley M. Human Krüppel-like factor 8: A CACCC-box binding protein that associates with CtBP and represses transcription. Nucleic Acids Research. 2000;**28**:1955-1962

[99] Wang X, Zhao J. KLF8 transcription factor participates in oncogenic transformation. Oncogene. 2007;**26**:456-461

[100] Yan Q, Zhang W, Wu Y, Wu M, Zhang M, Shi X, et al. KLF8 promotes tumorigenesis, invasion and metastasis of colorectal cancer cells by transcriptional activation of FHL2. Oncotarget. 2015;**6**:25402-25417

[101] Yi X, Zai H, Long X, Wang X, Li W, Li Y. Krüppel-like factor 8 induces epithelial-to-mesenchymal transition and promotes invasion of pancreatic cancer cells through transcriptional activation of four and a half LIM-only protein 2. Oncology Letters. 2017;**14**:4883-4889

[102] Liu N, Wang Y, Zhou Y, Pang H, Zhou J, Qian P, et al. Krüppel-like factor 8 involved in hypoxia promotes the invasion and metastasis of gastric cancer via epithelial to mesenchymal transition. Oncology Reports. 2014;**32**:2397-2404

[103] Møller M, Strand SH, Mundbjerg K, Liang G, Gill I, Haldrup C, et al. Heterogeneous patterns of DNA methylation-based field effects in histologically normal prostate tissue from cancer patients. Scientific Reports. 2017;**7**:40636

[104] Martin KM, Metcalfe JC, Kemp PR. Expression of Klf9 and Klf13 in mouse development. Mechanisms of Development. 2001;**103**:149-151

[105] Morita M, Kobayashi A, Yamashita T, Shimanuki T, Nakajima O, Takahashi S, et al. Functional analysis of basic transcription element binding protein by gene targeting technology. Molecular and Cellular Biology. 2003;**23**:2489-2500

[106] Hu F, Knoedler J, Denver RJ. KrüPpel-like factor 9 enhances thyroid hormone receptor? Autoinduction in tadpole brain in vivo, increasing tissue sensitivity to thyroid hormone and accelerating metamorphosis. Frontiers in Endocrinology. DOI: 10.3389/conf.fendo.2011.03.00021

[107] Knoedler JR, Subramani A, Denver RJ. The Krüppel-like factor 9 cistrome in mouse hippocampal neurons reveals predominant transcriptional repression via proximal promoter binding. BMC Genomics. 2017;**18**:299

[108] Simmen RCM, Pabona JMP, Velarde MC, Simmons C, Rahal O, Simmen FA. The emerging role of Krüppel-like factors in endocrine-responsive cancers of female reproductive tissues. The Journal of Endocrinology. 2010;**204**:223-231

[109] Simmen RCM, Eason RR, McQuown JR, Linz AL, Kang T-J, Chatman L Jr, et al. Subfertility, uterine hypoplasia, and partial progesterone resistance in mice lacking the Krüppel-like factor 9/basic transcription element-binding protein-1 (Bteb1) gene. The Journal of Biological Chemistry. 2004;**279**:29286-29294

[110] Simmen FA, Xiao R, Velarde MC, Nicholson RD, Bowman MT, Fujii-Kuriyama Y, et al. Dysregulation of intestinal crypt cell proliferation and villus cell migration in mice lacking Krüppel-like factor 9. American Journal of Physiology: Gastrointestinal and Liver Physiology. 2007;**292**:G1757-G1G69

[111] BennettLB, SchnabelJL, KelchenJM, Taylor KH, Guo J, Arthur GL, et al. DNA hypermethylation accompanied by transcriptional repression in follicular lymphoma. Genes, Chromosomes and Cancer. 2009;**48**:828-841

[112] Kang L, Lai M-D. BTEB/KLF9 and its transcriptional regulation. Hereditas. 2007;**29**:515-522

[113] Subramaniam M, Hawse JR, Rajamannan NM, Ingle JN, Spelsberg TC. Functional role of KLF10 in multiple disease processes. BioFactors (Oxford, England). 2010;**36**:8-18

[114] Spittau B, Krieglstein K. Klf10 and Klf11 as mediators of TGF-beta superfamily signaling. Cell and Tissue Research. 2012;**347**:65-72

[115] Yajima S, Lammers CH, Lee SH, Hara Y, Mizuno K, Mouradian MM. Cloning and characterization of murine glial cell-derived neurotrophic factor inducible transcription factor (MGIF). The Journal of Neuroscience: The Official Journal of the Society for Neuroscience. 1997;**17**:8657-8666

[116] Subramaniam M, Harris SA, Oursler MJ, Rasmussen K, Riggs BL, Spelsberg TC. Identification of a novel TGF-beta-regulated gene encoding a putative zinc finger protein in human osteoblasts. Nucleic Acids Research. 1995;**23**:4907-4912

[117] Liu Y, Peng W-Q, Guo Y-Y, Liu Y, Tang Q-Q, Guo L. Krüppel-like factor 10 (KLF10) is transactivated by the transcription factor C/EBPβ and involved in early 3T3-L1 preadipocyte differentiation. The Journal of Biological Chemistry. 2018;**293**(36):14012-14021

[118] Yerges LM, Klei L, Cauley JA, Roeder K, Kammerer CM, Ensrud KE, et al. Candidate gene analysis of femoral neck trabecular and cortical volumetric bone mineral density in older men. Journal of Bone and Mineral Research: The Official Journal of the American

Society for Bone and Mineral Research. 2010;**25**:330-338

[119] Subramaniam M, Gorny G, Johnsen SA, Monroe DG, Evans GL, Fraser DG, et al. TIEG1 null mouse-derived osteoblasts are defective in mineralization and in support of osteoclast differentiation in vitro. Molecular and Cellular Biology. 2005;**25**:1191-1199

[120] Bensamoun SF, Tsubone T, Subramaniam M, Hawse JR, Boumediene E, Spelsberg TC, et al. Age-dependent changes in the mechanical properties of tail tendons in TGF-beta inducible early gene-1 knockout mice. Journal of Applied Physiology (Bethesda, MD: 1985). 2006;**101**:1419-24

[121] Rajamannan NM, Subramaniam M, Abraham TP, Vasile VC, Ackerman MJ, Monroe DG, et al. TGFbeta inducible early gene-1 (TIEG1) and cardiac hypertrophy: Discovery and characterization of a novel signaling pathway. Journal of Cellular Biochemistry. 2007;**100**:315-325

[122] Cao Z, Wara AK, Icli B, Sun X, Packard RRS, Esen F, et al. Krüppel-like factor KLF10 targets transforming growth factor-beta1 to regulate CD4(+) CD25(−) T cells and T regulatory cells. The Journal of Biological Chemistry. 2009;**284**:24914-24924

[123] Chang VHS, Chu P-Y, Peng S-L, Mao T-L, Shan Y-S, Hsu C-F, et al. Krüppel-like factor 10 expression as a prognostic indicator for pancreatic adenocarcinoma. The American Journal of Pathology. 2012;**181**:423-430

[124] Asano H, Li XS, Stamatoyannopoulos G. FKLF, a novel Krüppel-like factor that activates human embryonic and fetal β-like globin genes. Molecular and Cellular Biology. 1999;**19**:3571-3579

[125] D'Souza UM, Lammers C-H, Hwang CK, Yajima S, Mouradian MM. Developmental expression of the zinc finger transcription factor DRRF (dopamine receptor regulating factor). Mechanisms of Development. 2002;**110**:197-201

[126] Song C-Z, Gavriilidis G, Asano H, Stamatoyannopoulos G. Functional study of transcription factor KLF11 by targeted gene inactivation. Blood Cells, Molecules, and Diseases. 2005;**34**:53-59

[127] Cook T, Gebelein B, Mesa K, Mladek A, Urrutia R. Molecular cloning and characterization of TIEG2 reveals a new subfamily of transforming growth factor-beta-inducible Sp1-like zinc finger-encoding genes involved in the regulation of cell growth. The Journal of Biological Chemistry. 1998;**273**:25929-25936

[128] Cook T, Gebelein B, Belal M, Mesa K, Urrutia R. Three conserved transcriptional repressor domains are a defining feature of the TIEG subfamily of Sp1-like zinc finger proteins. The Journal of Biological Chemistry. 1999;**274**:29500-29504

[129] Scohy S, Gabant P, Van Reeth T, Hertveldt V, Dreze PL, Van Vooren P, et al. Identification of KLF13 and KLF14 (SP6), novel members of the SP/XKLF transcription factor family. Genomics. 2000;**70**:93-101

[130] Mathison A, Grzenda A, Lomberk G, Velez G, Buttar N, Tietz P, et al. Role for Krüppel-like transcription factor 11 in mesenchymal cell function and fibrosis. PLoS One. 2013;**8**:e75311

[131] Buck A, Buchholz M, Wagner M, Adler G, Gress T, Ellenrieder V. The tumor suppressor KLF11 mediates a novel mechanism in transforming growth factor beta-induced growth inhibition that is inactivated in pancreatic cancer. Molecular Cancer Research. 2006;**4**:861-872

[132] Lomberk G, Grzenda A, Mathison A, Escande C, Zhang J-S, Calvo E, et al. Krüppel-like factor 11 regulates the expression of metabolic genes via an evolutionarily conserved protein-interaction domain functionally disrupted in maturity onset diabetes of the young. The Journal of Biological Chemistry. 2013;**288**:17745-17758

[133] Neve B, Fernandez-Zapico ME, Ashkenazi-Katalan V, Dina C, Hamid YH, Joly E, et al. Role of transcription factor KLF11 and its diabetes-associated gene variants in pancreatic beta cell function. Proceedings of the National Academy of Sciences of the United States of America. 2005;**102**:4807-4812

[134] Gutiérrez-Aguilar R, Froguel P, Hamid YH, Benmezroua Y, Jørgensen T, Borch-Johnsen K, et al. Genetic analysis of Krüppel-like zinc finger 11 variants in 5864 Danish individuals: Potential effect on insulin resistance and modified signal transducer and activator of transcription-3 binding by promoter variant -1659G>C. The Journal of Clinical Endocrinology and Metabolism. 2008;**93**:3128-3135

[135] Fernandez-Zapico ME, van Velkinburgh JC, Gutiérrez-Aguilar R, Neve B, Froguel P, Urrutia R, et al. MODY7 gene, KLF11, is a novel p300-dependent regulator of Pdx-1 (MODY4) transcription in pancreatic islet beta cells. The Journal of Biological Chemistry. 2009;**284**:36482-36490

[136] Ellenrieder V, Buck A, Harth A, Jungert K, Buchholz M, Adler G, et al. KLF11 mediates a critical mechanism in TGF-beta signaling that is inactivated by Erk-MAPK in pancreatic cancer cells. Gastroenterology. 2004;**127**:607-620

[137] Fernandez-Zapico ME, Mladek A, Ellenrieder V, Folch-Puy E, Miller L, Urrutia R. An mSin3A interaction domain links the transcriptional activity of KLF11 with its role in growth regulation. The EMBO Journal. 2003;**22**:4748-4758

[138] Potapova A, Hasemeier B, Römermann D, Metzig K, Göhring G, Schlegelberger B, et al. Epigenetic inactivation of tumour suppressor gene KLF11 in myelodysplastic syndromes*. European Journal of Haematology. 2010;**84**:298-303

[139] Wang G, Li X, Tian W, Wang Y, Wu D, Sun Z, et al. Promoter DNA methylation is associated with KLF11 expression in epithelial ovarian cancer. Genes, Chromosomes and Cancer. 2015;**54**:453-462

[140] Imhof A, Schuierer M, Werner O, Moser M, Roth C, Bauer R, et al. Transcriptional regulation of the AP-2alpha promoter by BTEB-1 and AP-2rep, a novel wt-1/egr-related zinc finger repressor. Molecular and Cellular Biology. 1999;**19**:194-204

[141] Godin-Heymann N, Brabetz S, Murillo MM, Saponaro M, Santos CR, Lobley A, et al. Tumour-suppression function of KLF12 through regulation of anoikis. Oncogene. 2016;**35**:3324-3334

[142] Nakamura Y, Migita T, Hosoda F, Okada N, Gotoh M, Arai Y, et al. Krüppel-like factor 12 plays a significant role in poorly differentiated gastric cancer progression. International Journal of Cancer. 2009;**125**:1859-1867

[143] Rozenblum E, Vahteristo P, Sandberg T, Bergthorsson JT, Syrjakoski K, Weaver D, et al. A genomic map of a 6-Mb region at 13q21-q22 implicated in cancer development: Identification and characterization of candidate genes. Human Genetics. 2002;**110**:111-121

[144] Zhang J-S, Moncrieffe MC, Kaczynski J, Ellenrieder V, Prendergast FG, Urrutia R. A conserved α-helical motif mediates the interaction of Sp1-like transcriptional repressors

with the corepressor mSin3A. Molecular and Cellular Biology. 2001;**21**:5041-5049

[145] Suda S, Rai T, Sohara E, Sasaki S, Uchida S. Postnatal expression of KLF12 in the inner medullary collecting ducts of kidney and its trans-activation of UT-A1 urea transporter promoter. Biochemical and Biophysical Research Communications. 2006;**344**:246-252

[146] Yuan J, Kang J, Yang M. Long non-coding RNA ELF3-antisense RNA 1 promotes osteosarcoma cell proliferation by upregulating Krüppel-like factor 12 potentially via methylation of the microRNA-205 gene. Oncology Letters. 2020;**19**:2475-2480

[147] Gordon AR, Outram SV, Keramatipour M, Goddard CA, Colledge WH, Metcalfe JC, et al. Splenomegaly and modified erythropoiesis in KLF13−/− mice. The Journal of Biological Chemistry. 2008;**283**:11897-11904

[148] Zhou M, McPherson L, Feng D, Song A, Dong C, Lyu SC, et al. Krüppel-like transcription factor 13 regulates T lymphocyte survival in vivo. Journal of Immunology. 2007;**178**:5496-5504

[149] Outram SV, Gordon AR, Hager-Theodorides AL, Metcalfe J, Crompton T, Kemp P. KLF13 influences multiple stages of both B and T cell development. Cell Cycle (Georgetown, Tex). 2008;**7**:2047-55

[150] Song A, Patel A, Thamatrakoln K, Liu C, Feng D, Clayberger C, et al. Functional domains and DNA-binding sequences of RFLAT-1/KLF13, a Krüppel-like transcription factor of activated T lymphocytes. The Journal of Biological Chemistry. 2002;**277**:30055-30065

[151] Huang B, Ahn Y-T, McPherson L, Clayberger C, Krensky AM. Interaction of PRP4 with Krüppel-like factor 13 regulates CCL5 transcription. The Journal of Immunology. 2007;**178**:7081-7087

[152] Lavallée G, Andelfinger G, Nadeau M, Lefebvre C, Nemer G, Horb ME, et al. The Krüppel-like transcription factor KLF13 is a novel regulator of heart development. The EMBO Journal. 2006;**25**:5201-5213

[153] Darwich R, Li W, Yamak A, Komati H, Andelfinger G, Sun K, et al. KLF13 is a genetic modifier of the Holt-Oram syndrome gene TBX5. Human Molecular Genetics. 2017;**26**:942-954

[154] Wu R, Yun Q, Zhang J, Bao J. Downregulation of KLF13 through DNMT1-mediated hypermethylation promotes glioma cell proliferation and invasion. OncoTargets and Therapy. 2019;**12**:1509-1520

[155] Koh I-U, Lee H-J, Hwang J-Y, Choi N-H, Lee S. Obesity-related CpG methylation (cg07814318) of Krüppel-like factor-13 (KLF13) gene with childhood obesity and its cis-methylation quantitative loci. Scientific Reports. 2017;**7**:45368

[156] Parker-Katiraee L, Carson AR, Yamada T, Arnaud P, Feil R, Abu-Amero SN, et al. Identification of the imprinted KLF14 transcription factor undergoing human-specific accelerated evolution. PLoS Genetics. 2007;**3**:e65

[157] Stacey SN, Sulem P, Masson G, Gudjonsson SA, Thorleifsson G, Jakobsdottir M, et al. New common variants affecting susceptibility to basal cell carcinoma. Nature Genetics. 2009;**41**:909-914

[158] Small KS, Hedman ÅK, Grundberg E, Nica AC, Thorleifsson G, Kong A, et al. Identification of an imprinted master trans regulator at the KLF14 locus related to multiple metabolic phenotypes. Nature Genetics. 2011;**43**:561-564

[159] Guo Y, Fan Y, Zhang J, Lomberk GA, Zhou Z, Sun L, et al. Perhexiline activates KLF14 and reduces atherosclerosis by modulating ApoA-I production. The Journal of Clinical Investigation. 2015;**125**:3819-3830

[160] Wei X, Yang R, Wang C, Jian X, Li L, Liu H, et al. A novel role for the Krüppel-like factor 14 on macrophage inflammatory response and atherosclerosis development. Cardiovascular Pathology. 2017;**27**:1-8

[161] Hu W, Lu H, Zhang J, Fan Y, Chang Z, Liang W, et al. Krüppel-like factor 14, a coronary artery disease associated transcription factor, inhibits endothelial inflammation via NF-kappaB signaling pathway. Atherosclerosis. 2018;**278**:39-48

[162] de Assuncao TM, Lomberk G, Cao S, Yaqoob U, Mathison A, Simonetto DA, et al. New role for Krüppel-like factor 14 as a transcriptional activator involved in the generation of signaling lipids. The Journal of Biological Chemistry. 2014;**289**:15798-15809

[163] Uchida S, Tanaka Y, Ito H, Saitoh-Ohara F, Inazawa J, Yokoyama KK, et al. Transcriptional regulation of the CLC-K1 promoter by myc-associated zinc finger protein and kidney-enriched Krüppel-like factor, a novel zinc finger repressor. Molecular and Cellular Biology. 2000;**20**:7319-7331

[164] Wang B, Haldar SM, Lu Y, Ibrahim OA, Fisch S, Gray S, et al. The Krüppel-like factor KLF15 inhibits connective tissue growth factor (CTGF) expression in cardiac fibroblasts. Journal of Molecular and Cellular Cardiology. 2008;**45**:193-197

[165] Fisch S, Gray S, Heymans S, Haldar SM, Wang B, Pfister O, et al. Krüppel-like factor 15 is a regulator of cardiomyocyte hypertrophy. Proceedings of the National Academy of Sciences of the United States of America. 2007;**104**:7074-7079

[166] Pepin ME, Ha C-M, Crossman DK, Litovsky SH, Varambally S, Barchue JP, et al. Genome-wide DNA methylation encodes cardiac transcriptional reprogramming in human ischemic heart failure. Laboratory Investigation: A Journal of Technical Methods and Pathology. 2019;**99**:371-86

[167] Gray S, Wang B, Orihuela Y, Hong E-G, Fisch S, Haldar S, et al. Regulation of gluconeogenesis by Krüppel-like factor 15. Cell Metabolism. 2007;**5**:305-312

[168] Haldar SM, Jeyaraj D, Anand P, Zhu H, Lu Y, Prosdocimo DA, et al. Krüppel-like factor 15 regulates skeletal muscle lipid flux and exercise adaptation. Proceedings of the National Academy of Sciences. 2012;**109**:6739-6744

[169] Takeuchi Y, Yahagi N, Aita Y, Murayama Y, Sawada Y, Piao X, et al. KLF15 enables rapid switching between lipogenesis and gluconeogenesis during fasting. Cell Reports. 2016;**16**:2373-2386

[170] Jeyaraj D, Scheer FAJL, Ripperger JA, Haldar SM, Lu Y, Prosdocimo DA, et al. Klf15 orchestrates circadian nitrogen homeostasis. Cell Metabolism. 2012;**15**:311-323

[171] Wang J, Galvao J, Beach KM, Luo W, Urrutia RA, Goldberg JL, et al. Novel roles and mechanism for Krüppel-like factor 16 (KLF16) regulation of neurite outgrowth and ephrin receptor A5 (EphA5) expression in retinal ganglion cells. The Journal of Biological Chemistry. 2016;**291**:18084-18095

[172] Daftary GS, Lomberk GA, Buttar NS, Allen TW, Grzenda A, Zhang J, et al. Detailed structural-functional analysis of the Krüppel-like factor 16 (KLF16) transcription factor reveals novel mechanisms

for silencing Sp/KLF sites involved in metabolism and endocrinology. The Journal of Biological Chemistry. 2012;**287**:7010-7025

[173] van Vliet J, Crofts LA, Quinlan KG, Czolij R, Perkins AC, Crossley M. Human KLF17 is a new member of the Sp/KLF family of transcription factors. Genomics. 2006;**87**:474-482

[174] Yan W, Burns KH, Ma L, Matzuk MM. Identification of Zfp393, a germ cell-specific gene encoding a novel zinc finger protein. Mechanisms of Development. 2002;**118**:233-239

[175] Zhou S, Tang X, Tang F. Krüppel-like factor 17, a novel tumor suppressor: Its low expression is involved in cancer metastasis. Tumour Biology. 2016;**37**:1505-1513

[176] Gumireddy K, Li A, Gimotty PA, Klein-Szanto AJ, Showe LC, Katsaros D, et al. KLF17 is a negative regulator of epithelial-mesenchymal transition and metastasis in breast cancer. Nature Cell Biology. 2009;**11**:1297-1304

[177] Ali A, Bhatti MZ, Shah AS, Duong HQ, Alkreathy HM, Mohammad SF, et al. Tumor-suppressive p53 signaling empowers metastatic inhibitor KLF17-dependent transcription to overcome tumorigenesis in non-small cell lung cancer. The Journal of Biological Chemistry. 2015;**290**:21336-21351

[178] Cai X-D, Che L, Lin J-X, Huang S, Li J, Liu X-Y, et al. Krüppel-like factor 17 inhibits urokinase plasminogen activator gene expression to suppress cell invasion through the Src/p38/MAPK signaling pathway in human lung adenocarcionma. Oncotarget. 2017;**8**:38743-38754

[179] Ali A, Zhang P, Liangfang Y, Wenshe S, Wang H, Lin X, et al. KLF17 empowers TGF-beta/Smad signaling by targeting Smad3-dependent pathway to suppress tumor growth and metastasis during cancer progression. Cell Death & Disease. 2015;**6**:e1681

[180] Dong P, Kaneuchi M, Xiong Y, Cao L, Cai M, Liu X, et al. Identification of KLF17 as a novel epithelial to mesenchymal transition inducer via direct activation of TWIST1 in endometrioid endometrial cancer. Carcinogenesis. 2014;**35**:760-768

[181] Ali A, Ielciu I, Alkreathy HM, Khan AA. KLF17 attenuates estrogen receptor alpha-mediated signaling by impeding ERalpha function on chromatin and determines response to endocrine therapy. Biochimica et Biophysica Acta. 1859;**2016**:883-895

[182] Pei J, Grishin NV. A new family of predicted Krüppel-like factor genes and pseudogenes in placental mammals. PLoS One. 2013;**8**:e81109

[183] Shankman LS, Gomez D, Cherepanova OA, Salmon M, Alencar GF, Haskins RM, et al. KLF4 dependent phenotypic modulation of SMCs plays a key role in atherosclerotic plaque pathogenesis. Nature Medicine. 2015;**21**:628-637

[184] Wirka RC, Wagh D, Paik DT, Pjanic M, Nguyen T, Miller CL, et al. Atheroprotective roles of smooth muscle cell phenotypic modulation and the TCF21 disease gene as revealed by single-cell analysis. Nature Medicine. 2019;**25**:1280-1289

[185] Buenrostro JD, Wu B, Chang HY, Greenleaf WJ. ATAC-seq: A method for assaying chromatin accessibility genome-wide. Current Protocols in Molecular Biology. 2015;**109**:21.9.1-9

[186] Shashikant T, Ettensohn CA. Genome-wide analysis of chromatin accessibility using ATAC-seq. Methods in Cell Biology. 2019;**151**:219-235

[187] Liu J, Jia G. Methylation modifications in eukaryotic messenger

RNA. Journal of Genetics and Genomics. 2014;**41**:21-33

[188] Dominissini D, Moshitch-Moshkovitz S, Schwartz S, Salmon-Divon M, Ungar L, Osenberg S, et al. Topology of the human and mouse m^6A RNA methylomes revealed by m^6A-seq. Nature. 2012;**485**:201-206

[189] Fu Y, Dominissini D, Rechavi G, He C. Gene expression regulation mediated through reversible m^6A RNA methylation. Nature Reviews. Genetics. 2014;**15**:293

[190] Linder B, Grozhik AV, Olarerin-George AO, Meydan C, Mason CE, Jaffrey SR. Single-nucleotide-resolution mapping of m^6A and m^6Am throughout the transcriptome. Nature Methods. 2015;**12**:767-772

[191] Niu Y, Zhao X, Wu Y-S, Li M-M, Wang X-J, Yang Y-G. N6-methyl-adenosine (m^6A) in RNA: An old modification with A novel epigenetic function. Genomics, Proteomics & Bioinformatics. 2013;**11**:8-17

[192] Zhang Z, Chen L-Q, Zhao Y-L, Yang C-G, Roundtree IA, Zhang Z, et al. Single-base mapping of m^6A by an antibody-independent method. Science Advances. 2019;5:eaax0250

Chapter 6

Circular RNAs and Its Biological Functions in Health and Disease

Atiye Seda Yar Saglam, Ebru Alp and Hacer Ilke Onen

Abstract

Circular RNAs (circRNAs) belong to the family of long noncoding RNAs (lncRNA) that, unlike linear RNAs, are characterized by a covalently closed circular RNA structure lacking 5′ cap and 3′ poly-adenylated tails. circRNAs have a role in epigenetic regulation of downstream targets. circRNAs play a crucial role in regulating gene and protein expressions by acting as a microRNA (miRNA) sponge and RNA binding protein (RBP) sponge and interact with proteins to affect cell behavior. circRNA expression profiles differ between physiological and pathological states. Moreover, the expression patterns of circRNAs exhibit differences in a tissue-specific manner. Although investigations on circRNAs have been exploding nowadays, yet only a limited number of circRNAs are identified. Furthermore, further researches are needed to shed light on their functions and targets. Therefore, circRNAs are becoming vital as potential biomarkers that may be used for the diagnosis and treatment of diseases. In this chapter, we review the current advancement of cirRNAs with regard to their biogenesis, biological functions, gene regulatory mechanisms, and implications in human diseases and summarize the recent studies on circRNAs as potential diagnostic and prognostic biomarkers based on existing knowledge.

Keywords: circular RNAs, cardiovascular diseases, neurological disorders, immune regulation, cancer

1. Introduction

The ENCyclopedia Of DNA Elements (ENCODE) project reported that noncoding RNAs (ncRNAs) unexpectedly consist of more than 70% of the human genome [1]. After the data released by ENCODE project consortium, numerous studies have focused on the identification and function of these transcripts [2]. ncRNAs can be a group based on their different characteristic features [3]. Long noncoding RNAs (lncRNAs) are subclass of ncRNAs that have been recently proved to have a role in physiological and pathological processes [4]. lncRNAs are >200 nucleotides long, divergent class of RNA transcripts that coordinate expression of protein-coding genes. Yet, they have a lack of ability to encode proteins [5]. Circular RNAs (circRNAs) are a special subtype of lncRNAs [6]. circRNAs are characterized by a single-stranded covalently closed loop structure with neither a 5′ cap nor a 3′ poly (A) tail [7]. Due to their circular structure, circRNAs are more stable than the linear mRNA counterpart and not susceptible to RNA exonuclease cleavage [6, 7]. The presence of circRNA was first demonstrated in the cytoplasm of eukaryotic cells in

1979 [8]. It was thought that circRNAs were a by-product formed during splicing mechanism in the first year [9]. Numerous circRNAs have been predicted with the technical developments in high-throughput RNA sequencing (RNA-seq) and methodological innovations in bioinformatics. The presence and function of the predicted circRNAs in different tissues and cell lines are widely studied nowadays. After the determination of their role in the control of gene expression, circRNAs have gained great attention by researchers in this field. In this chapter, we will focus on circRNAs and their biological functions in health and disease.

2. Biogenesis of circRNAs

According to the gene structure they contain, circRNAs can be divided into three groups: exonic circRNA (ecircRNA), circRNAs from introns (ciRNAs), and exon-intron circRNA (elciRNA) [10]. To date, many studies have shown that circRNAs mainly emerged during pre-mRNA splicing process of protein coding genes. Unlike canonical mRNA splicing mechanism, down-stream donor splice site is covalently joined with an upstream acceptor splice sites during circRNA formation. This splicing mechanism is called "back-splicing" [7]. The back-splicing mechanism is depicted in **Figure 1**. circRNAs can also be formed through the hybridization of complementary inverted sequences (such as human Alu repeats) in introns [10]. If Alu sequences are located in different introns of the same gene, this leads to

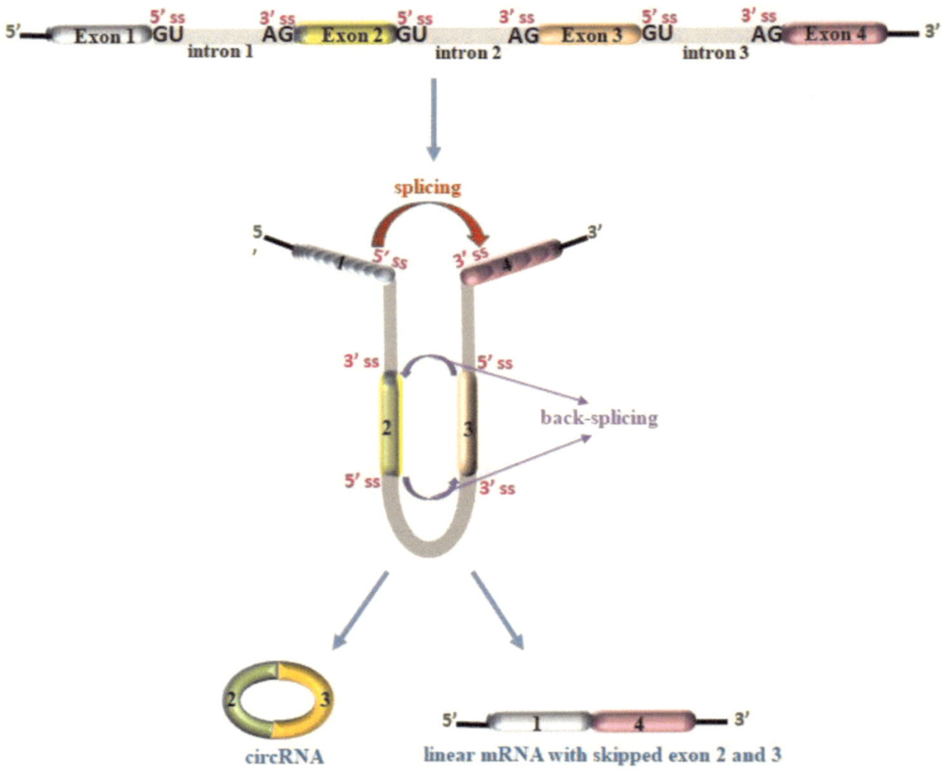

Figure 1.
Schematic illustration of the circRNA formation by back splicing mechanism. Unlike canonical mRNA splicing mechanism, the 3′ splice donor site of exon 1 binds to the 5′ splice acceptor site of exon 4 during circRNA formation. The back-splicing results in a circRNA including exon 2 and 3 and linear mRNA with skipped exon 2 and 3. ss, splice site.

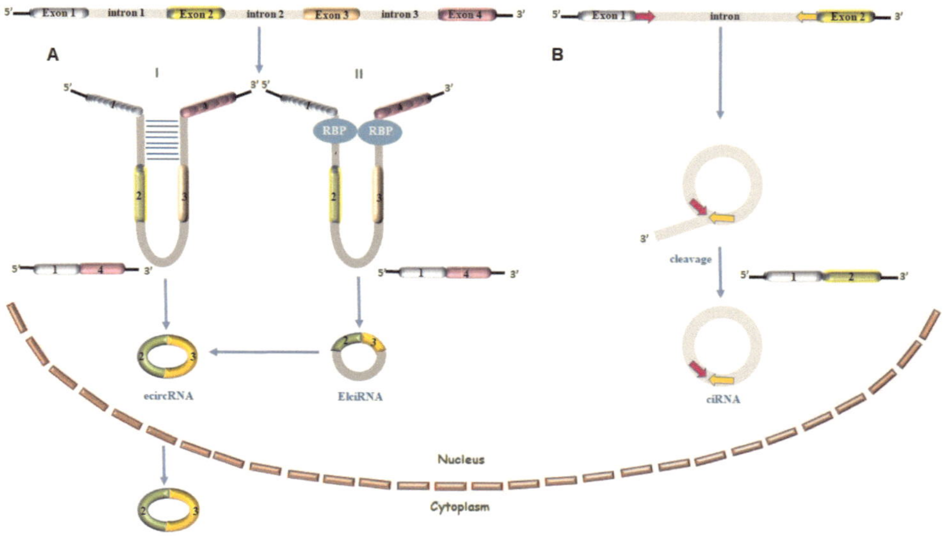

Figure 2.
Possible model for the formation of structurally different circRNA. (AI) Intron-driven circularization: circRNAs may form by hybridization of the introns with inverted repeats or Alu sequence. (AII) RBP-driven circularization: RBPs bind to specific sequence in introns that bring the exons close together and trigger the circularization. At the end of these two ways, elciRNAs or ecircRNAs are generated. Only ecircRNAs can be transported to cytoplasm. (B) Formation of ciRNA. The ciRNAs are generated from intron lariat by splicing reaction. Purple arrow represents 7-nt GU-rich element. Yellow arrow represents 11-nt C-rich element near to the branchpoint site.

generate circRNAs, which contains multiple exons [7]. Exon circularization is facilitated by cis-acting inverted repeat sequences as well as by trans-acting RNA-binding proteins (RBPs), which interact with unique sequences in introns [11, 12]. ecircRNA or elciRNA formation is promoted by either cis-acting elements or trans-acting factors as in **Figure 2A**. ecircRNAs can also be formed from elciRNAs by removal of intronic sequences [13]. Apart from the other circRNAs, ecircRNAs are transported into cytoplasm [14]. In human cells, the existence of ciRNAs is demonstrated by Zhang et al. [15]. ciRNAs are generated through a lariat-derived mechanism relying on mainly a consensus motif containing a 7-nt GU-rich element adjacent to the 5′ splice site and an 11-nt C-rich element adjacent to the branchpoint site. After cleavage of 3′ end, stable ciRNA is produced [15]. The predicted biogenesis of ciRNA is shown **Figure 2B**.

3. Gene regulation and biologic functions of circRNAs

The expression patterns of circRNAs are specific to the cell type or phase of development [16]. Although the all-biological functions of circRNAs are not entirely defined, some are well studied in the literature. Biological functions of circRNAs include micro RNA (miRNA) sponge, regulation of gene expression and properties of mRNA binding, scaffolding, and cellular translocation.

3.1 circRNAs can act as miRNA sponges

As a major component of gene regulators, competing/competetive endogenous RNA (ceRNA) contains a micro RNA response element (MRE-competitively binds miRNA) and can affect the regulatory functions of miRNAs [17]. Growing evidence has indicated that circRNAs can act as ceRNA or miRNA sponge molecules. Because of containing plenty of MRE, circRNAs can competitively bind to miRNAs (generally several copies of miRNA) and adsorb them like a sponge [18]. As a result, miRNAs can no longer act on their target mRNA [19]. Therefore, circRNAs can regulate the gene expression and also give rise to decreasing of the functional miRNA [17, 18]. Compared to other ceRNAs, circRNA binds more effectively to miRNAs. Therefore, they are also called "super sponge" [20]. The most characteristic miRNA sponge "antisense to the cerebellar degeneration related protein 1 transcript" (CDR1as)/ciRS-7 includes approximately 70 conserved binding sites for miRNA-7 (miR-7) and forms a complex with Argonaute (AGO) proteins [21]. CDR1as-miR-7 complex co-localizes in the cytoplasm and supresses degradation of miR-7-target mRNAs [17]. Interestingly, circRNA has been reported that it is also displayed to be abundant in exosomes in serum [16]. Therefore, Li et al. suggested that sorting of circRNAs to exosomes was regulated by altering levels of associated miRNA in producer cells [16, 22]. In addition, researchers have found that CDR1as including exosomes inhibit miR-7-induced growth in recipient cells [22]. Testes-specific circRNA/circSry [the circular transcript of sex determining region Y (Sry) gene] can also serve as a sponge for miRNA-138 [23]. It contains 16 MREs of miRNA-138 and regulates the expression of miR-138-target genes, functioning similar to CDR1as [17]. Additionaly, many other circRNAs have been identified as miRNA sponges such as hsa_circ_001569, heart-related circRNA (HRCR), itchy E3 ubiquitin protein ligase circRNA (circITCH), forkhead box O3 circRNA (circ-foxo3), homeodomain interacting protein kinase 3 circRNA (circHIPK3), mitochondrial tRNA translation optimization 1 circRNA (circMTO1), zinc finger protein 609 circRNA (cirZNF609), and baculoviral IAP repeat containing 6 circRNA (circBIRC6) [24]. Among them, cirITCH regulates the expression of ITCH by acting as a sponge for miR-214, miR-17, and miR-7 [22].

Apart from common phenomenon, in some cases, the binding of circRNAs to miRNAs may not always lead to inhibition of miRNAs. Linearization and AGO2-mediated cleavage of CDR1as can occur when CDR1as interacts with miR-671. Thus, bound miR-7 is released from CDR1as [25]. On the other hand, in spite of the role of circRNA in gene regulation as a classical sponge effect, some recent studies have revealed that the number of circRNAs with miRNA sponge property is limited. Besides inhibition effect, it has been regarded that the interaction between circRNAs and miRNA is also related to sorting, storage and localization of miRNA [18].

3.2 circRNAs regulate gene expression and interact with protein

In addition to their miRNA sponge function, circRNAs can also act as sponges for other components as RBPs. There are many proteins known as RBP such as AGO protein, RNA quaking, muscleblind (MBL) protein, RNA polymerase II (Pol II), eukaryotic initiation factor 4A-III [26]. RBPs bind specific sequences to their target genes and control all stages of mRNA lifecycle including splicing, nuclear export, stability and subcellular localisation [27]. A number of circRNAs contain a large amount of binding sites for a single or multiple RNA-binding proteins. For example, circRNA protein sponge derived from the *MBL* locus includes binding sites of mbl protein. Thus, mbl is prevented from binding to other targets. In a parallel study, circular RNA of polyadenylate-binding nuclear protein 1 (circPABPN1) derived from PABPN1 gene binds to HuR (enhance PABPN1 translation) and prohibits its binding to PABPN1 mRNA [28].

circRNAs also inhibits parental gene transcription in target genes via invading RNA binding sites. Strongly binding of circRNA, derived from sepallata3 (SEP3) gene, to its cognate DNA locus blocks the binding of its linear isoform to the cognate DNA. The formation of R-loop (RNA:DNA hybrid) gives rise to termination of SEP3 gene transcription [13]. Moreover, circRNAs can be paired with DNA to generate DNA-RNA triple helixes. Therefore, this pairing may affect DNA replication [29].

Some ciRNAs (e.g., ci-ankrd52, ci-sirt7) and eIciRNAs (e.g., circEIF3J, circPAIP2) regulate the transcription of their parental genes. eIciRNAs can regulate parental gene transcription in a cis-acting manner [18, 23, 24]. Recent studies have been indicated that nuclear elciRNAs (localized to the promoter of their parental genes) interact with U1 small nuclear ribonucleoproteins (snRNPs) and RNA pol II and promote the transcription initiation of their parental genes [28, 30]. For example, eukaryotic translation initiation factor 3J circRNA (circEIF3J) and poly(A)-binding protein-interacting protein 2 circRNA (circPAIP2) have been suggested to have cis-regulatory effects on parental genes and promote transcription of EIF3J and PAIP2. This cis-regulatory effect occurs by binding of its circRNAs to Pol II, U1 snRNP, and their parental gene promoters [18, 25]. When transcription is initiated, the production of eIciRNA can be increased so that this phenomenon generates a positive feedback loop [23]. Moreover, they have a function as positive regulators through their interactions with the elongating Pol II complex [24, 25, 31]. In addition, exonic circular antisense noncoding RNA in the INK4 (a family of cyclin-dependent kinase inhibitors) locus (circANRIL/cANRIL) reduces ANRIL that inhibits transcription of INK4/ARF gene by binding to the Polycomb Gene (PcG) complex. Thus, cANRIL regulates the transcription of INK4/ARF [32].

3.3 Cellular translocation properties

Some circRNAs may affect nuclear translocation of other proteins to nucleus and regulation of gene transcription. For example, CircAmotl1 may increase nuclear translocation of signal transducer and activator of transcription 3 (STAT3) to regulate the expression of mitosis-related genes [24].

Another capability of circRNAs is to ensure that cellular proteins remain in their natural cellular position. It has been reported that circAmotl1 can enhance stability of c-myc by maintaining its nuclear retention and increase its binding affinity to several promoters. Therefore, it upregulates c-myc targets such as hypoxia inducible factor-1 alpha (HIF-1α), cell division cycle 25A (Cdc25a), ETS Like-1 (ELK-1) [24]. In another example, cytoplasmic circ-foxo3 interacts with differentiation-1 (ID-1), HIF-1α, focal adhesion kinase (FAK), the transcription factor E2 (E2F1) and prevents their translocation from cytoplasm to other location [13].

3.4 Scaffolding properties

circRNAs also serve as scaffolding in the assembly of protein complexes [13]. It has been reported that circ-foxo3 acts as an adaptor to bridge between cyclin-dependent kinase 2 (CDK2) and CDK inhibitor p21 (cyclin-dependent kinase inhibitor 1A). This interaction (circ-foxo3/CDK2/p21) inhibits cell-cycle progression within G1 to S-phase transition [18, 24]. However, downregulation of circ-foxo3 leads to the release of CDK2 from p21 and CDK2 phosphorylates cyclin E and cyclin A for cell cycle progression. On the other hand, the circ-foxo3 connects the murine double minute 2 (MDM2) to tumor protein p53 (p53), and induces the degradation of p53 by ubiquitination. However, circ-foxo3 weakly interacts with foxo3 and suppresses foxo3 from MDM2-mediated polyubiquitination and proteasome degradation [30].

3.5 mRNA binding properties

Most circRNAs are capable of interaction with mRNAs. It has been reported that they can be able to regulate the stability of mRNAs as well. In addition to its miRNA sponge function, CDR1as is also proposed to form a duplex structure with CDR1 mRNA and stabilizes it. Similarly, stabilization of mature intercellular adhesion molecule 1 mRNAs in macrophages was found to be facilitated by RasGEF domain family member 1B circRNA (circRasGEF1B) [18].

3.6 The effect of circRNA as a translator

Recent studies have shown that some circRNAs can be entered to translational process in spite of considering noncoding RNA [33, 34]. A limited number of studies have indicated the potential protein coding properties of circRNAs until now but the translational efficiency might be low [33]. circRNA containing internal ribosomal entry site (IRES) and open reading frame can be translated into protein or polypeptide. In eukaryotes, IRES is an alternative way of initiating translation, independent of 5′ cap structure and 3′ poly (A) tail recognition [19]. It has been demonstrated that the 40S subunit of the eukaryotic ribosome can interact with circRNA-containing IRES and then begin translation in *in vivo* and *in vitro* experiments. It has been shown that zinc finger protein 609 circRNA (circZNF609) can be translated into a novel ZNF609 protein isoform and potential function during myogenesis. Another study indicated that novel proteins have been translated from F-box and WD repeat domain containing 7 circRNA (circFBXW7) and SNF2 histone linker PHD ring helicase circRNA (circSHPRH) in glioblastoma cell lines. A new isoform protein encoded by circFBXW7 with open reading frame was found to inhibit glioma cell growth [18, 34].

A recent study reported that N6-methyladenosine (m6A), a most common base modification of RNA, can promote the protein translation of circRNA in human cells, even if one m6A motif can initiate circRNA translation [18, 19, 29]. m6A-driven circRNA translation is prevalent, and several endogenous circRNAs have the potential for translation and regulatory role in a cell against enviromental factors [29].

3.7 The effect of circRNAs on splicing

Recent studies have shown that there is a competition between backsplicing and linear splicing. Thus, the biogenesis of circRNAs leads to loss of protein-coding mRNA levels and inhibits parental gene expression [13]. On the other hand, the level of circRNA is negatively correlated to the splicing efficiency of certain genes due to the competition between linear splicing and circRNA biogenesis [30].

4. circRNAs in cardiovascular diseases

Cardiovascular disease (CVD) is one of the most important health problems. It causes most of the deaths worldwide [35]. According to recent studies, a number of circRNAs may play a significant role during development of CVD or pathological conditions such as cardiac hypertrophy, coronary artery disease, atherosclerotic vascular disease, cardiomyopathy, cardiac fibrosis, heart failure (HF), ischemia, and myocardial infarction (MI) [36–38]. However, in development of heart disease, the regulatory mechanisms and functional importance of several circRNAs are not clear [38]. circRNAs are also concentrated in body fluids such as seminal fluid,

circRNA	Possible target	CVD	Biological function or description	Ref
circANRIL (ex 5–7)	PES-1	AS and CAD	Impairs pre-rRNA maturation and ribosome biogenesis and increases nucleolar stress and apoptosis	[32, 47]
circANRIL (ex 4–6)		ASVD	Neighboring gene regulation such as INK4a	[32]
Hsa_circ_0003575	miR-199-3p, mir-9-5p, mir-377-3p, and miR-141-3p	AS	Regulates endothelial cell proliferation and angiogenesis acting as a miRNA sponge	[48]
Hsa_circ_0010729	mir-186	AS and CHD	Regulates vascular endothelial cell proliferation and apoptosis via targeting miR-186 and HIF-1α axis	[49]
circACTA2	miR-548f-5p	AS and CHD	Maintains contractile phenotype of VSMC Mediates NRG-1-ICD regulation of α-SMA expression in HASMCs	[50]
circWDR77	mir-124	AS	Regulates VSMC proliferation and migration via targeting miR-124 and FGF2 Inhibits the expression of SM22a and STIM1 by acting as a miRNA sponge	[51]
circ-SATB2	mir-939	AS	Inhibits the expression of SM22a and STIM1 by acting as a miRNA sponge Regulates cell phenotypic differentiation, proliferation, apoptosis, and migration in VSMC	[52]
circR-284	miR-221	AS and carotid plaque rupture	Reduces the proliferation of VSMCs by circR-284/mir-221/$p27^{Kipi}$ axis Upregulated circR-284:miR-221 ratio in the early stage of carotid plaque rupture	[53, 54]
hsa_circ_0124644		CAD	Potential biomarker of coronary artery disease	[55]
hsa_-circ_0001879		CAD	Significant upregulated expression levels in CAD patients	[56]
hsa_circ_0004104			Dysregulation of atherosclerosis-related genes by overexpression of hsa_circ_0004104	[56]
CDR1as	miR-7a	MI	Upregulates the expression of PARP and SP1 acting as a miRNA sponge and promotes apoptosis	[57]
MICRA		Acute MI, HF, and LVD	Potential biomarker of left ventricular dysfunction in the patients with acute MI	[58, 59]
MFACR	miR-652-3p	MI	Upregulates apoptosis and mitochondrial fission	[60]

circRNA	Possible target	CVD	Biological function or description	Ref
HRCR	mir-223	HF and cardiac hypertrophy	Increases the expression of ARC by acting as a miRNA sponge. Suppresses cardiac hypertrophy	[61]
circ-081881	mir-548	Acute MI	Positively regulates PPARγ acting as a miRNA sponge	[62]
circRNA-010567	miR-141	MI	May mediate fibrosis-associated protein resection	[63]
circNCX1	miR-133a-3p	Ischemic myocardial injury	Promotes cardiomyocyte apoptosis by acting as a miRNA sponge and increased in response to ROS	[64]
circAmotl1	AKT and PDK	Cardiac repair and cardiomyopathy	Facilitates the nuclear translocation of AKT and PDK1 Improves survival and decreases apoptosis	[45]
circTTN		DCM	Dysregulated in disease model	[65, 66]
circRyr2		Cardiomyopathy		[65]
circZNF609	miR-615-5p and miR-150-5p	Hypertension and CAD	Inhibits cell proliferation, migration, and tube formation and promotes cell apoptosis Acts as a miRNA sponge and leads to upregulation of MEF2A expression	[67]
Hsa-circ-0005870	hsa-miR-619-5p, hsa-miR-5095, hsa-miR-1273 g-3p, and hsa-miR-5096	Hypertension	Downregulated in hypertension patients	[68]
rno_circRNA_016002		Hypertension	Upregulated in hypertensive rat strains compared to normotensive rats	[69]
hsa_circ_0014243	hsa-miR-10a-5p	EH	Crucial role in the genesis and development of EH and presents a certain diagnostic capability for EH	[70]
hsa_circ_0037911	miR-637	EH	Upregulated in hypertension patients	[65, 71]
hsa_circ_0126991		EH	May serve as a stable biomarker for early diagnosis of EH	[72]
circ-foxo3	ID-1, E2F1, FAK, and HIF-1α	Cardiac senescence	Interacts with ID-1, E2F1, FAK, and HIF-1α and induces cellular senescence in aging hearts	[46]

CVD, cardiovascular disease; AS, atherosclerosis; ASVD, atherosclerotic vascular disease; CHD, coronary heart disease; CAD, coronary artery disease; MI, myocardial infarction; HF, heart failure; LVD, left ventricular dysfunction; EH, essential hypertension; DCM, dilated cardiomyopathy; VSMC, vascular smooth muscle cell; ROS, reactive oxygen species; ANRIL, antisense noncoding RNA in the INK4 locus; PES1, pescadillo homologue 1; ACTA2, actin alpha 2; WDR77, WD repeat domain 77; STIM1, stromal interaction molecule 1; SATB2, special AT-rich sequence-binding protein 2; CDR1, cerebellar degeneration-related protein 1; MICRA, myocardial infarction associated circRNA; MFCAR, mitochondrial fission and apoptosis-related circRNA; HRCR, heart-related circRNA; ARC, apoptosis repressor with CARD domain; NCX1, sodium/ calcium exchanger 1; AMOTL1, angiomotin like 1; AKT, protein kinase B; PDK, phosphoinositide-dependent protein kinase; TTN, titin; RYR2, ryanodine receptor 2; Foxo3, forkhead box O3; ID1, inhibitor of DNA binding 1; E2F1, E2F transcription Factor 1; FAK, focal adhesion kinase; HIF-1α, hypoxia inducible factor-1; FGF2, fibroblast growth factor2; PARP, poly ADP-ribose polymerase; and MEF2A, myocyte enhancer factor 2A.

Table 1.
Summary of identified circRNA in the cardiovascular disease.

saliva, and blood. Thus, their potential usage as clinical biomarkers may be possible in the future [39].

Some heart specific RNA-splicing regulators are also important players for heart development. One of the RNA-splicing regulators is RBM20 that is required for splicing of cardiac-related genes such as titin [38]. Its mutation leads to exon retention in the region of I-band and results in larger titin isoforms [30]. According to RNA-seq researches in tissues from dilated cardiomyopathy (DCM) and hypertrophic cardiomyopathy, 80 different circRNAs are derived from the titin gene (TTN) [30, 40].

circANRIL is generated as an antisense transcript from the INK4A/ARF gene locus by alternative splicing [36]. SNPs localized within chromosome 9p21 are likely to affect the INK4/ARF locus. These SNPs can regulate ANRIL splicing and may lead to circANRIL production [39]. Interestingly, there is an association between 9p21 SNPs and the susceptibility to atherosclerosis [41]. circANRIL is also implicated in the pathogenesis of atherosclerosis [42]. In another study, Burd et al. suggested that 9p21 SNPs affect the coordination of ANRIL expression and splicing via interaction of different PcG complexes. Furthermore, PcG complexes are targeted to the INK4/ARF locus and that leads to inhibition of INK4/ARF transcription. Moreover, they also indicated that their study is the first to provide evidence for relationship between circRNA and atherosclerotic vascular disease (ASVD) [32].

circANRIL also disrupts exonuclease-mediated pre-rRNA processing and ribosome biogenesis by binding to pescadillo homologue 1 (PES1). This leads to nuclear stress and p53 activation in cells [38, 39]. Therefore, it supresses cell proliferation and inhibits apoptosis in vascular smooth muscle cells and macrophages. Consequenly, circANRIL acts as a protective factor against atherosclerosis [41, 43]. On the other hand, it has been indicated that a novel circular RNA product of ANRIL, cANRIL (exon4-6) also regulates the expression of INK4/ARF [32].

In addition, circRNA serves as a protein scaffold such as circAmotl1 in cardiac dysfunction [35, 43]. circAmotl1 facilitates phosphorylation of protein kinase B (AKT) and nuclear translocation of pAKT by forming ternary complexes with AKT and phosphoinositide-dependent protein kinase (PDK) [43–45]. Zeng et al. have suggested that pAKT translocation may be responsible for protection of heart cells from cardiomyopathy caused by doxorubicin [45].

circ-foxo3 is another circRNA described to may have a role in the cardiovascular diseases. Stress-related proteins (HIF-1α and FAK) and senescence-related proteins [inhibitor of DNA-binding protein (ID1) and E2F1] are arrested in cytoplasm by circ-foxo3. Therefore, circ-foxo3 prevents translocation of these proteins into the nucleus. As a consequence, this mechanism promotes cardiac senecence through ectopic expression of circ-foxo3 [41, 46]. Besides these functions, circRNAs are reported to also show their effects as miRNA sponge in cardiovascular diseases. circRNAs and their function in cardiovascular diseases are indicated in **Table 1**.

Although there is limited number of studies until today, CVD-related studies for circRNA are in progress. Therefore, it is still required the identification of circRNA as candidate biomarkers for CVDs. Moreover, biologic functions of circRNA in vascular endothelial cell and heart tissue should be validated in further studies.

5. circRNAs in neurological disorders

Recent studies have shown that circRNAs are plentifully expressed in normal neuronal cells [73–75]. They may be found abundantly in neuronal cells for several reasons: (i) brain contains more host genes of circRNA such as neuronal genes, which play roles in neurogenesis, neuronal development, and neuronal differentiation [11, 74], (ii) the expression levels of circRNAs are higher in brain than other tissues [75, 76], (iii) due to the slow division rates of neurons, circRNAs may accumulate more in the brain than other tissues [77], (iv) neuronal genes contain long introns (>10 kb) with inverted repeat sequences, thereby simplifying formation of circRNAs [10], and (v) circRNAs due to the absence of 5′ and 3′ ends result in greater stability than linear RNAs, leading to a relatively longer half-life [78]. The half-life of circRNAs is approximately 20 h, compared with corresponding linear isoforms (no more than 8 h) [79].

The latest studies have shown that circRNAs could attenuate cell senescence and cell survival and may be involved in the regulation of aging and age-related neurological diseases [80–82]. Thus, circRNAs are expected to be new potential biomarkers and target for aging and age-related neurological diseases (**Table 2**). These studies have suggested that circRNAs may play an important role in pathological mammalian brain function, which is implicated in disorders in central nervous system (CNS) including Alzheimer's disease (AD), Parkinson's disease (PD), neuropsychiatric disorders, prion disease, and inflammatory neuropathy.

CDR1as, a circRNA, is highly plentiful and specifically expressed in the mammalian brain [85]. Some studies have indicated that ciRS-7 contains multiple anti-miR-7 sequences. This suggests that ciRS-7 may function as a sponge to sequester the normal functions of miR-7 [57, 95–97]. ciRS-7 can regulate the stability of mRNA targets in the brain by binding to miR-7 [78, 85]. Besides, ciRS-7 can interact

CircRNA	Target	Neurological disease	Possible mechanisms	Ref
ciRS-7	miR-7	AD	ciRS-7 is reduced in AD, and miR-7 can downregulate AD relevant targets, such as ubiquitin conjugating enzyme UBE2A, which play an essential role in the clearance of amyloid peptides	[83, 84]
circSry	miR-138	AD	mir-138 participate in learning and memory ability and is increased in AD, and it promotes tau phosphorylation by targeting the RARA/GCK-3β pathway	[85–87]
ciRS-7	miR-7	PD	miR-7 may downregulate α-synuclein expression, promotes the degradation of α-synuclein mRNA levels, and protects cells against oxidative stress	[88]
ciRS-7	miR-7 miR-671	Neuropsychiatric disorders	miRNA deregulation and affects brain function	[78, 89, 90]
ciRS-7	miR-7	Prion disease	Prion protein PrPc can upregulate expression of ciRS-7	[91, 92]
hsa-circRNA 2149	—	Inflammatory neuropathy	Hsa-circRNA 2149 has been detected in CD19+ leukocytes	[53]
circSry	miR-138 circRNA100783	Inflammatory neuropathy	miR-138 can balance the ratio of Th1 and Th2 via suppressing the function of RTF3 CircRNA100783 may be involved in chronic CD28-associated CD8 (+) T cell aging	[93, 94]

AD, Alzheimer's disease; PD, Parkinson disease; UBE2A, ubiquitin conjugating enzyme E2 A; RARA/GCK-3β, retinoic acid receptor alpha/glycogen synthase kinase-3β; and RTF3, runt-related transcription factor 3.

Table 2.
Functional mechanism of cirRNAs in neurological disease.

with multiple protein subunits, thus acting as "scaffolding" for RBPs [7, 98]. Thereby, it facilitates the interaction by potentially increasing the stability of the circRNA transcripts. Due to its multiple functions in brain, researchers have suggested that ciRS-7 can be a potential biomarker for neurodegenerative disorders including AD and PD [83].

5.1 circRNA in Alzheimer's disease

Alzheimer's disease is a chronic neurological disease. Lukiw et al. showed that the expression level of ciRS-7 is decreased in hippocampal CA1 region in sporadic AD [83]. Functional deficiency of ciRS-7 can lead to upregulation of miR-7 in AD brain and may cause the downregulation of several AD-relevant mRNA targets, including the ubiquitin conjugating enzyme E2A (UBE2A) [83, 84, 99, 100]. This autophagic protein, UBE2A, is a central effector in the ubiquitination cycle. UBE2A is crucial for clearing amyloid peptides via phagocytosis and contributes to amyloidogenesis [99]. In contrast to the previous studies, Shi et al. have shown that ciRS-7 promotes the degradation of amyloid precursor protein (APP) and beta-secretase 1 (BACE1) in an nuclear factor kappa beta (NF-κB)-dependent manner [101]. Hence, future studies are needed to reveal ciRS-7 function/functions and its exact role in AD pathology.

CircSry can serve as a miRNA sponge in neural cells. CircSry inhibits miR-138 [53, 85], which is a potential molecular regulator of human memory function [102]. CircSry has multiple binding sites for miR-138 and promotes tau phosphorylation by targeting the "retinoic acid receptor alpha/glycogen synthase kinase-3β" (RARA/GCK-3β) pathway [86]. Some studies have indicated that miR-138 influences learning and memory abilities by regulating acyl protein thioesterase 1 [87, 102]. Therefore, association of circSry and miR-138 in AD should be further investigated.

5.2 circRNA in Parkinson's disease

Parkinson disease, progressive age-related neurodegenerative disorder, is characterized by the loss of dopaminergic neurons in the substantia nigra pars compacta [103, 104]. To date, five genes have been determined to cause PD, such as α-synuclein (SNCA), parkin, dj-1, PTEN-induced kinase 1 (pink1), and leucine-rich repeat kinase 2 (lrrk2) [105]. SNCA is the key player in the pathogenesis of PD based on neuropathologic, genetic, and cellular evidence [106]. The over-expression and aggregation of SNCA, a target gene of miR-7, is considered as a distinctive marker in PD [107, 108]. miR-7 has been proposed to play a role in PD by reducing expression of SNCA [88]. ciRS-7 plays a protective role by inhibiting miR-7 that directly regulates the expression of SNCA [109]. miR-7 alleviates SNCA expression dose-dependently and induces the degradation of SNCA mRNA levels [88]. These results suggest that ciRS-7 serves as a miR-7 sponge *in vitro*. Furthermore, the silencing of ciRS-7 increases miR-7 activity and decreases the expression of miR-7 target genes [110]. In addition, circSNCA, another circRNA, can sponge miR-7, thereby regulating expression of SNCA, resulting in decreased autophagy and increased apoptosis in cells [111]. These findings are in concordance with the results of a study, which showed that autophagy can prevent PD [112], and that of the other study, which demonstrated that apoptosis is related to PD [113].

5.3 circRNA and inflammatory neuropathy

circRNAs may participate in inflammatory reactions that induce neuropathy. Some circRNAs may affect immune responses due to the fact that they contain virus miRNA binding sites. For instance, hsa-circRNA 2149 contains 13 unique, head to tail spanning reads. Researchers discovered hsa-circRNA 2149 in CD19+ leukocytes, but not CD341 leukocytes or neutrophils. On the other hand, circRNA100783 may be involved in chronic CD28-related CD8(+) T cell aging and for this reason could be a novel biomarker for this conditions [93]. Furthermore, circSry, another circRNA, can repress miR-138 activity, which could balance T helper 1 (Th1) and T helper 2 (Th2) expressions through suppressing the function of runt-related transcription factor 3 (RUNX3) [94].

5.4 circRNA and prion diseases

Most prion diseases are infectious via transmissible particles composed of prion protein in scrapie (PrPSc), an isomer of noninfectious cellular prion protein (PrPc). Studies have discovered that ciRS-7 expression is induced by PrPc overexpression [91, 92]. ciRS-7 may suppress miR-7 activity and therefore ciRS-7 may be involved in the prion disease pathogenesis.

5.5 circRNA and neuropsychiatric disorders

Apart from in brain tumors, ciRS-7 may also play a role in neuropsychiatric disorders. Increased miR-7 levels have been determined in neuropsychiatric disorders, serving as a proof for ciRS-7-mediated deregulation of dendritic spine density via a miR-7-SHANK3 (SH3 and multiple ankyrin repeat domains 3) axis [89, 90]. In recent study, Piwecka et al. showed that ciRS-7 knockout mice display behavioural phenotypes related to neuropsychiatric disorders. Deleting of ciRS-7 locus in mice leads to synaptic transmission function disorder and unusual neuropsychiatric-like behavior [78]. Other than miR-7, ciRS-7 also has a binding site to miR-671, which is deregulated in all brain regions in ciRS-7 deficient mice; however, the direction of changes was opposite. It is designated that the binding site on ciRS-7 is completely complementary to miR-671, and the interaction of these two molecules could lead to AGO-mediated ciRS-7 slicing and miR-671 deterioration. On the contrary, the binding sites on ciRS-7 are partial complementaries to miR-7. For this reason, it is likely that circRNAs can serve as a platform to store and transport certain miRNAs [78, 89, 90].

Currently, circRNA studies in the CNS are in progress. To date, there is a limited number of circRNA identified in neurological disorders. Moreover, previous studies mainly focus on ciRS-7 function. Therefore, it is still needed to identify candidate circRNAs as a potential biomarker in neurological disease. In addition, their functional properties in neuronal cells should be also validated in further studies.

6. The role of circRNAs in immune regulation

Although many circRNAs are under survey, their roles in autoimmune diseases remain incomprehensible, and there are insufficient data to determine their exact role of circRNAs in such diseases [24, 114].

The connection between miRNAs and immunity has been well-studied, which has led to the hypothesis that circRNAs may contribute to immune regulation by interacting with miRNAs. In particular, due to their abilities to serve as miRNA and protein sponges, they can regulate gene expression and encode proteins. Therefore, circRNAs can participate in the development and progression of different immune responses and immune diseases [23, 24, 114]. On the basis of the current studies, the majority of circRNAs defined in autoimmune diseases are ecircRNAs, and a few are

Disease	CircRNA	Regulation	miRNA sponge targets	Potential functions	Ref
SLE	Hsa_circ_102584	↑	miR-766-3p miR-762 miR-412-3p let-7i-3p miR-431-3P	It may be improved as novel noninvasive biomarkers for SLE	[117]
	Hsa_circ_400011	↑	miR-296-3p miR-146b-3p miR-181d-3p miR-504-3p		
	Hsa_circ_101471	↑	miR-328-5p miR-136-5p miR-665 miR-486-3p miR-601		
	Hsa_circ_100226	↓	miR-30b-3p miR-138-5p miR-145-3p miR-24-3p miR-620 miR-875-3p		
	CDR1as/ciRS-7	↓	—	It functions as the miR-7 sponge to increase expression of PTEN and restricts hyper-responsiveness of B cells	[100, 118]
RA	Hsa_circ_104871	↑	—	It serves as potential biomarkers for diagnosis and performs severity or pathological course of RA	[119]
	Hsa_circ_003524	↑	—		
	Hsa_circ_101873	↑	—		
	Hsa_circ_103047	↑	—		
	Hsa_circ_0057980	↓	miR-181d	It functions as the miR-181d sponge to suppress the development of RA	[86, 120]
	Hsa_circ_0088088	↓	miR-16	It functions as the miR-16 sponge to suppress the development of RA	[121–123]
	Hsa_circ_0001045	↑	miR-30a	It functions as the miR-30a sponge to promote the biogenesis of RA	
MS	Hsa_circ_0005402	↓	—	It can be improved as MS biomarkers	[124]
	Hsa_circ_0035560	↓	—	It arranges negatively the biogenesis of MS	
	GSDMB ecircRNA	↑	miR-1275	It functions as the miR-1275 and miR-149 sponges to induce MS	[124, 125]
			miR-149	Both circRNAs are derived from the ANXA2	
PBC	Hsa_circ_402458	↑	miR-522-3p	It may be appropriate for PBC diagnosis	[121, 126]
			miR-943	It functions as the miR-522 and miR-943 sponges to counter chronic	[127]

Disease	CircRNA	Regulation	miRNA sponge targets	Potential functions	Ref
				inflammation and aberrant TGF-β signalling of PBC	
SCID	Circ-CDC42BPA	↑	—	It disrupts transduction of B cell signalling to induce formation of SCID	[128, 129]
	Circ-TNFRSF11A	↑	—	It attends in the SCID-mediated alteration of different signalling pathways	[128, 130]
WAS	Circ-ROBO1	↑	—	It activates the pathogenesis of WAS	[128, 131]
	Circ-CDC42BPA	↑	—	It disrupts transduction of B cell signalling to induce formation of WAS	[128]

MS, multiple sclerosis; PBC, primary biliary cirrhosis; RA, rheumatoid arthritis; SCID, severe combined immunodeficiency disease; SLE, systemic lupus erythematosus; and WAS, Wiskott-Aldrich syndrome.

Table 3.
circRNAs are associated with immune diseases.

ciRNAs and elciRNAs [23, 24, 114–116]. The circRNAs identified to date, their functions, and roles in immunological diseases are shown in **Table 3**.

It will be important in future studies to determine biological functions of circRNAs in immune cells. circRNAs may serve as both potential biomarkers and immune regulators [23, 24, 114–116]. Hence, it may be helpful to improve our understanding of the molecular biological basis of autoimmune diseases.

7. circRNAs in cancer

Cancer is one of the most common causes of death in worldwide. As stated in world cancer report (2014), 10 million people of the world develop all types of cancer each year. Moreover, over 6 million patients around the world die from this disease annually [132]. Unfortunately, the number of patients diagnosed with cancer is increasing and is estimated to increase in future in worldwide [133, 134]. Even if, a functional improvement in the treatment approach is established, and new therapeutic strategies are still needed for therapy of cancer. Therefore, the identification of the altered pathways and gene transcripts has been the subject of researches recently. miRNAs have a role in gene regulation and affect various molecular biological processes such as cell growth, development, differentiation, proliferation, and cell death [135]. As circRNAs interact with miRNAs and then influence the mRNA expression levels of target genes, the identification of circRNA-miRNA-mRNA network has become the objective of cancer researches.

There are numerous investigations on circRNAs and their functions in cancer as compared with other diseases. To date, most of the studies have focused on miRNA sponge function of circRNAs. miRNAs have been classified depending on the effect of miRNAs on downstream target/targets [136]. miRNAs can act as oncogenes or tumor suppressors during carcinogenesis [137]. Likewise, circRNAs are also named according to their behaviour during tumorigenesis. While some circRNAs contribute to tumor progression and metastasis, the others suppress oncogenesis.

DOI: http://dx.doi.org/10.5772/intechopen.88764

CircRNA	circRNA expression status	Target miRNA	miRNA expression status	Target mRNA/ signaling pathway	mRNA expression status	Main findings of the studies	Ref
hsa_circ_0001946	↓	hsa-miR-7-5p, hsa-miR-671-5p, hsa-miR-1270, hsa-miR-3156-5p	↑	NER signaling pathway	Activated	Compared to pairs of adjacent nontumor tissues, expression of hsa_circ_0001946 is downregulated in 43 NSCLC tissues There was a decrease in hsa_circ_0001946 expression on the cisplatin-resistant A549/CDDP cells compared with the parental A549 cells	[138]
	↑	miR-135a-5p	↓	SIRT1	↑	Compared to pairs of adjacent nontumor tissues, expression of circ_0001946 is upregulated in 72 lung adenocarcinoma tissues The circ_0001946 expression is upregulated in the four lung adenocarcinoma cell lines compared with the nonmalignant human lung epithelial cell line The increase in circ_0001946 expression in tumor samples is an independent prognostic factor for the patients with lung adenocarcinoma as well as advanced TNM stages	[139]
circAGFG1	↑	miR-203	↓	ZNF281	↑	Compared to pairs of adjacent nontumor tissues, expression of circAGFG1 is upregulated in 20 NSCLC tissues circAGFG1 enhances ZNF281-mediated migration and proliferation of NSCLC	[140]
hsa_circRNA_102984 (circPTPRA)	↓	miR-96-5p	↑	RASSF8/ e-cadherin	↑	Compared to pairs of adjacent nontumor tissues, expression of hsa_circRNA_102984 (circPTPRA) is downregulated in 34 NSCLC tissues circPTPRA acts as a miR-96-5p sponge, and it leads to upregulation of RASSF8 levels in both *in vitro* and H23 xenograft model	[141]
circ_0020123	↑	miR-488e3p	↓	ADAM9	↑	Compared to pairs of adjacent nontumor tissues, expression of circ_0020123 is upregulated in 55 NSCLC tissues	[142]

CircRNA	circRNA expression status	Target miRNA	miRNA expression status	Target mRNA/ signaling pathway	mRNA expression status	Main findings of the studies	Ref
						The circ_0020123 expression is upregulated in the four NSCLC cell lines compared with the nonmalignant human bronchial epithelial cells The increase in circ_0020123 expression in tumor samples has been correlated with short overall survival rate in NSCLC patients	
	↑	miR-144	↓	ZEB1 EZH2	↑	Compared to pairs of adjacent nontumor tissues, expression of hsa_circ_0020123 is upregulated in 80 NSCLC tissues Upregulation of hsa_circ_0020123 expression in tumor samples has been correlated with short overall survival in NSCLC patients The hsa_circ_0020123 expression is upregulated in the six lung cancer cell lines	[143]
circVANGL1	↑	miR-195	↓	Bcl2	↑	Compared to pairs of adjacent nontumor tissues, expression of circVANGL1 is upregulated in 95 NSCLC tissues The circVANGL1 expression is upregulated in the five NSCLC cell lines compared with the nonmalignant human bronchial epithelial cells Upregulation of circVANGL1 expression leads to higher stage, bigger tumor size, and shorter overall survival in NSCLC patients	[144]
hsa_circRNA_102231 (hsa_circ_0046263) (named as circP4HB)	↑	miR-133a-5p	↓	Vimentin	↑	Compared to pairs of adjacent nontumor tissues, expression of circP4HB is upregulated in 80 NSCLC tissues Upregulation of circP4HB expression leads to higher metastatic capacity and shorter survival in NSCLC patients	[145]

CircRNA	circRNA expression status	Target miRNA	miRNA expression status	Target mRNA/ signaling pathway	mRNA expression status	Main findings of the studies	Ref
circ_0026134	↑	miR-1256 miR-1287	↓ ↓	TCTN1 and GAGE1	↑	Compared to pairs of adjacent nontumor tissues, expression of Circ_0026134 is upregulated in 30 NSCLC tissues The Circ_0026134 expression is upregulated in the four NSCLC cell lines compared with the nonmalignant human bronchial epithelial cells	[146]
Circ-FOXM1 (hsa_circ_0025033)	↑	miR-1304-5p	↓	PPDPF and MACC1	↑	Compared to pairs of adjacent nontumor tissues, expression of Circ-FOXM1 is upregulated in 80 NSCLC tissues The Circ-FOXM1 expression is upregulated in the four NSCLC cell lines compared with the nonmalignant human bronchial epithelial cells The increase in circ-FOXM1 expression in tumor samples was correlated with short overall survival rate in NSCLC patients	[147]
circ_0003645	↑	miR-1179	↓	TMEM14A	↑	Compared to pairs of adjacent nontumor tissues, expression of circ_0003645 is upregulated in 59 NSCLC tissues The circ_0003645 expression is upregulated in the four NSCLC cell lines compared with the nonmalignant human bronchial epithelial cells The increase in circ_0003645 expression in tumor samples is an independent prognostic factor for the patients with NSCLC as well as advanced TNM stages	[148]
hsa_circ_0002360	↑	hsa-mir-3620-5p	↓	PHF19	↑	Compared to pairs of adjacent nontumor tissues, expression of hsa_circ_0002360 is upregulated in 18 lung adenocarcinoma tissues	[149]
circRNA 100146	↑	miR-361-3p miR-615-5p	↓	SF3B3	↑	Compared to pairs of adjacent nontumor tissues, expression of circRNA 100146 is upregulated in 40 NSCLC tissues	[150]

CircRNA	circRNA expression status	Target miRNA	miRNA expression status	Target mRNA/ signaling pathway	mRNA expression status	Main findings of the studies	Ref
circFGFR3	↑	miR-22-3p	↓	Gal-1 Akt and Erk 1/2 signaling pathway	↑ Activated	Compared to pairs of adjacent nontumor tissues, expression of circFGFR3 is upregulated in 63 NSCLC tissues The increase in circFGFR3 expression in tumor samples is correlated with the poor prognosis of NSCLC patients	[151]
hsa_circ_0006427	↓	miR-6783-3p	↑	DKK1 Wnt/b-catenin signaling pathway	↑ Inactivated	Compared to pairs of adjacent nontumor tissues, expression of circ_0006427 is downregulated in 94 lung adenocarcinoma The circ_0006427 expression is downregulated in the four lung adenocarcinoma cell lines compared with the nonmalignant human lung epithelial cell line The decrease in circFGFR3 expression in tumor samples is correlated with the poor prognosis of lung adenocarcinoma patients	[152]
hsa_circ_0008305 circPTK2	↓	miR-429 miR-200b-3p	↑	TIF1γ	↓	circPTK2 has an important role in regulating TGF-β-induced EMT and tumor metastasis	[153]
hsa_circ_100395	↓	miR-1228	↑	TCF21		Compared to pairs of adjacent nontumor tissues, expression of hsa_circ_100395 is downregulated in 69 NSCLC The hsa_circ_100395 expression is downregulated in the six lung cancer cell lines compared with the the nonmalignant human bronchial epithelial cells Downregulation of hsa_circ_100395 expression in tumor samples is correlated with TNM stage and lymphoid node metastases	[154]
circ-BANP	↑	miR-503	↓	LARP1	↑	Compared to pairs of adjacent nontumor tissues, expression of circ-BANP is upregulated in 59 NSCLC The circ-BANP expression is upregulated in the four lung cancer cell lines compared with the nonmalignant human bronchial epithelial cells	[155]

DOI: http://dx.doi.org/10.5772/intechopen.88764

CircRNA	circRNA expression status	Target miRNA	miRNA expression status	Target mRNA/ signaling pathway	mRNA expression status	Main findings of the studies	Ref
						Upregulation of circ-BANP expression in tumor samples predicted lower Survival rate	
hsa_circRNA_103595 circMAN2B2	↑	miR-1275	↓	FOXK1	↑	Compared to pairs of adjacent nontumor tissues, expression of circMAN2B2 is upregulated in 41 NSCLC The circMAN2B2 expression is upregulated in the four lung cancer cell lines compared with the nonmalignant human lung epithelial cells	[156]
circ_0016760	↑	miR-1287	↓	GAGE1	↑	Compared to pairs of adjacent nontumor tissues, expression of circ_0016760 is upregulated in 83 NSCLC The circ_0016760 expression is upregulated in the four lung cancer cell lines compared with the nonmalignant human bronchial epithelial cells Upregulation of circ_0016760 expression in tumor samples predicted short overall survival in NSCLC patients	[157]

NER, nucleotide excision repair; NSCLC, nonsmall cell lung cancer; CDDP, cisplatin; SIRT1, sirtuin 1; AGFG1, ArfGAP with FG repeats 1; ZNF281, zinc finger protein 281; PTPRA, protein tyrosine phosphatase receptor type A; RASSF8, ras association domain family member 8; ADAM9, ADAM metallopeptidase domain 9; ZEB1, zinc finger E-box binding homeobox 1; EZH2, enhancer of zeste 2 polycomb repressive complex 2 subunit; VANGL1, VANGL planar cell polarity protein 1; BCL2, B-cell CLL/lymphoma 2; P4H1, prolyl 4-hydroxylase subunit beta; TCTN1, tectonic family member 1; GAGE1, G antigen 1; FOXM1, forkhead box M1; PPDPF, pancreatic progenitor cell differentiation and proliferation factor; MACC1, metastasis-associated in colon cancer 1; TMEM14A, transmembrane protein 14A; PHF19, PHD finger protein 19; SF3B3, splicing factor 3b subunit 3; FGFR3, fibroblast growth factor receptor 3; DKK, Dickkopf WNT signaling pathway inhibitor 1; PTK2, protein tyrosine kinase 2; TIF1γ, transcription intermediary factor 1-gamma; TGF-β, tumor growth factor beta; EMT, epithelial-mesenchymal transition; TCF21, transcription factor 21; BANP, BANP BTG3 associated nuclear protein; LARP1, La ribonucleoprotein domain family member 1; MAN2B2, mannosidase alpha class 2B member 2; FOXK1, forkhead box K1; and GAGE, G antigen 1.

Table 4.
The expression profile of circRNA-miRNA-mRNA network in lung cancer tissues.

CircRNA	circRNA expression status	Target miRNA	miRNA expression status	Target mRNA/ signaling pathway	mRNA expression status	Main findings of the studies	Ref
circ_0006528	↑	miR-7-5p	↓	Raf1	↑	Compared to adjacent nontumor tissues, expression of circ_0006528 is upregulated in BCa tissues The increase in circ_0006528 expression in tumor samples has been correlated with advanced TNM stage and poor prognosis	[158]
				MAPK/ERK signaling pathway	Activated		
circKIF4a (hsa_circ_0007255)	↑	miR-375	↓	KIF4A	↑	Compared to pairs of adjacent nontumor tissues, expression of circKIF4A is upregulated in 57 TNBC tissues circKIF4A expression increased in the five TNBC cell lines compared with the four NTNBC and nonmalignant breast epithelial cell line The increase in circKIF4A expression in tumor samples has been correlated with worse outcome of TNBC patients	[159]
hsa_circ_0004771	↑	miR-653	↓	ZEB2	↑	Compared to pairs of adjacent nontumor tissues, expression of hsa circ 0004771 is upregulated in BCa tissues hsa circ 0004771 expression increased in the five BCa cell lines compared with nonmalignant breast epithelial cell line The increase in hsa circ 0004771 expression in tumor samples has been correlated poorer survival prognosis	[160]
circTADA2A-E6	↓	miR-203a-3p	↑	SOCS3	↓	Compared to adjacent nontumor tissues, expression of Hsa circTADA2A-E6 is downregulated in TNBC tissues The decline in Hsa circTADA2A-E6 expression in tumor samples was associated with poor patient survival for TNBC	[161]
circAGFG1	↑	miR-195-5p	↓	CCNE1	↑	Compared to adjacent nontumor tissues, expression of circAGFG1 is upregulated in TNBC tissues circAGFG1 expression increased in the six TNBC cell lines compared with nonmalignant breast epithelial cell line The expression levels of circAGFG1 were reversely correlated with overall survival of patients with TNBC	[162]
hsa_circ_000479	↑	miR-4753 miR-6809	↓	BCL11A	↑	Compared to pairs of adjacent nontumor tissues, expression of circEPSTI1 is upregulated in 10 TNBC tissues	[163]

CircRNA	circRNA expression status	Target miRNA	miRNA expression status	Target mRNA/ signaling pathway	mRNA expression status	Main findings of the studies	Ref
						The increase in circEPSTI1 expression in tumor samples was positively correlated with tumor size, lymph node infiltration and TNM stage, and associated with poor prognosis	
hsa_circ_0008039	↑	miR-432-5p	↓	E2F3	↑	Compared to pairs of adjacent nontumor tissues, expression of hsa_circ_0008039 is upregulated in 38 TNBC tissues hsa_circ_0008039 expression increased in the six BCa cell lines compared with nonmalignant breast epithelial cell line	[164]
hsa_circ_0007534	↑	miR-593	↓	MUC19	↑	Compared to pairs of adjacent nontumor tissues, expression of hsa_circ_0007534 is upregulated in 40 BCa tissues hsa_circ_0007534 expression increased in the five BCa cell lines compared with nonmalignant breast epithelial cell line	[165]
circRNA-000911	↓	miR-449a	↑	Notch1 NF-κB pathway	↓ Activated	Compared to pairs of adjacent nontumor tissues, expression of circRNA-000911 is downregulated in 35 BCa tissues hsa_circRNA_000911 expression decreased in the six BCa cell lines compared with nonmalignant breast epithelial cell line	[166]
hsa_circ_0001846 circ-UBAP2	↑	miRNA-661	↓	MTA1	↑	Compared to pairs of adjacent nontumor tissues, expression of circ-UBAP2 is upregulated in 78 TNBC tissues circ-UBAP2 expression increased in TNBC cell lines compared with nonTNBC cell lines The increase in circ-UBAP2 expression in tumor samples has been correlated with reduced OS in TNBC patients	[167]
circRNA_0005505 circIRAK3	↑	miR-3607	↓	FOXC1	↑	Compared to pairs of adjacent nontumor tissues, expression of CircIRAK3 is upregulated in 35 BCa tissues CircIRAK3 expression increased in TNBC cell lines compared with normal mammary epithelial or ER-positive cell lines The increase in CircIRAK3 expression in tumor samples has been correlated worse recurrence-free survival in breast cancer patients	[168]
circ_0005230	↑	miR-618	↓	CBX8	↑	Compared to pairs of adjacent nontumor tissues, expression of circ_0005230 is upregulated in 76 BCa tissues	[169]

CircRNA	circRNA expression status	Target miRNA	miRNA expression status	Target mRNA/ signaling pathway	mRNA expression status	Main findings of the studies	Ref
						circ_0005230 expression increased in six BCa cell lines compared with nonmalignant mammary epithelial cell lines The increase in circ_0005230 expression in tumor samples has been correlated worse overall survival in breast cancer patients	
hsa_circ_0007294 circANKS1B	↑	miR-148a-3p miR-152-3p	↓	USF1 TGF-β1/Smad signalling	↑ Activated	Compared to pairs of adjacent nontumor tissues, expression of CircANKS1B is upregulated in 23 TNBC tissues CircANKS1B expression increased in TNBC cell lines compared with NTNBC cell lines The increase in CircANKS1B expression in tumor samples has been correlated worse overall survival in breast cancer patients	[170]
hsa_-circ_005239 circGFRA1	↑	miR-34a	↓	GFRA1	↑	Compared to pairs of adjacent nontumor tissues, expression of circGFRA1 is upregulated in 51 TNBC tissues The increase in circGFRA1 expression in tumor samples has been correlated short overall survival in TNBC patient circGFRA1 expression increased in TNBC cell lines compared with NTNBC cell lines	[171]

KIF4A, kinesin family member 4A; ZEB2, zinc finger E-box binding homeobox 2; CCNE1, cyclin E1; FOXC1, forkhead box C1; TNBC, triple negative breast cancer; NTNBC, nontriple negative breast cancer; Bca, Breast cancer; TADA2A, transcriptional adaptor 2A; SOCS3, suppressor of cytokine signaling 3; AGFG1, ArfGAP with FG repeats 1; EPSTI1, epithelial stromal interaction 1; BCL11A, B-cell CLL/lymphoma11A; E2F3, E2F transcription factor 3; MUC19, mucin 19; NOTCH1, notch receptor 1; NF-κB, nuclear factor Kappa beta; UBAP2, ubiquitin associated protein 2; MTA1, metastasis associated 1; IRAK3, interleukin 1 receptor associated kinase 3; CBX8, chromobox 8; ANKS1B, ankyrin repeat and sterile alpha motif domain containing 1B; USF1, upstream transcription factor 1; GFRA1, GDNF family receptor alpha 1; and TGF-β1, transforming growth factor beta 1.

Table 5.
The expression profile of circRNA-miRNA-mRNA network in breast cancer tissues.

Studies on altered expression of circRNAs in (lung and breast cancer) tumor samples are summarized in **Tables 4** and **5**. Moreover, in these selected studies, the circRNA-miRNA-mRNA interaction network is well defined.

By taking all studies together, circRNAs may be candidate surrogate molecular markers for cancer in different aspects, such as angiogenesis, metastasis, and drug resistance. Although to date some circRNA-miRNA-mRNA axis is predicted in cancer-associated pathways, the function and importance of dysregulated circRNAs still need to be supported in larger numbers of samples and patients, in various cancers.

8. Research databases of circRNA

With the increasing interest in circRNAs, comprehensive circRNA databases are required for prediction of circRNAs and their targets [172]. To evaluate and simplify the properties and interaction of various circRNAs with other RNAs from different aspects, numerous databases have been published (circlncRNAnet, starBase v2. 0, circBase, circRNABase, circ2Traits, nc2Cancer, DeepBase v2. 0, CircInteractome, TSCD, CIRCpedia, circRNADb, CircNet, CircR2Disease, circBank, and so on) [173]. Examples of circRNA databases and their usage in researches are shown.

- starBase v2. 0 determines miRNA-circRNA interactome and includes miRNA, mRNA, and lncRNA information [174].

- circ2Traits can be provided information about miRNA-circRNA interaction and its association with particular diseases [109].

- CircInteractome can be used in coupling the circRNA with related RNA-binding proteins [175].

- TSCD is helpful to describe tissue-specific circRNAs in mouse and human genomes [176].

- CIRCpedia includes reverse and variable splicing sites of circRNAs from individuals and mouse samples [177].

- circBank can be a resource to facilitate the research of function and regulation of circRNAs [178].

9. Conclusion

In summary, circRNAs, a new class of noncoding RNAs, are widely investigated by researchers due to their role in post transcriptional gene regulation. Recent studies have indicated their effects on the development of diverse diseases by acting as a miRNA sponge, RBP sponge, and transcriptional modulator or direct encoding proteins. Although the miRNA sponge function of circRNAs is currently investigated in the diseases, other mechanisms of circRNAs are still under investigation, and further studies are needed. After the interpretation of their function in disease pathogenesis, they may have a potential to become a drug target. Using circRNAs as biomarkers or therapeutic targets needs to be further investigated due to their complex roles. Based on these characteristics, circRNAs are likely to guide the development of new diagnostic and therapeutic strategies as well as prevention of diseases.

Conflict of interest

The authors declare no conflict of interest.

Author details

Atiye Seda Yar Saglam[1], Ebru Alp[2] and Hacer Ilke Onen[1*]

1 Department of Medical Biology, Faculty of Medicine, Gazi University, Ankara, Turkey

2 Department of Medical Biology, Faculty of Medicine, Giresun University, Giresun, Turkey

*Address all correspondence to: ilkeonen@yahoo.com; hionen@gazi.edu.tr

References

[1] ENCODE Project Consortium. An integrated encyclopedia of DNA elements in the human genome. Nature. 2012;**489**:57-74. DOI: 10.1038/nature11247

[2] Clark MB, Choudhary A, Smith MA, Taft RJ, Mattick JS. The dark matter rises: The expanding world of regulatory RNAs. Essays in Biochemistry. 2013;**54**:1-16. DOI: 10.1042/bse0540001

[3] St. Laurent G, Wahlestedt C, Kapranov P. The landscape of long noncoding RNA classification. Trends in Genetics. 2015;**31**:239-251. DOI: 10.1016/j.tig.2015.03.007

[4] Batista PJ, Chang HY. Long noncoding RNAs: Cellular address codes in development and disease. Cell. 2013;**152**:1298-1307. DOI: 10.1016/j.cell.2013.02.012

[5] Derrien T, Johnson R, Bussotti G, Tanzer A, Djebali S, Tilgner H, et al. The GENCODE v7 catalog of human long noncoding RNAs: Analysis of their gene structure, evolution, and expression. Genome Research. 2012;**22**:1775-1789. DOI: 10.1101/gr.132159.111

[6] Lasda E, Parker R. Circular RNAs: Diversity of form and function. RNA. 2014;**20**:1829-1842. DOI: 10.1261/rna.047126.114

[7] Jeck WR, Sharpless NE. Detecting and characterizing circular RNAs. Nature Biotechnology. 2014;**32**:453-461. DOI: 10.1038/nbt.2890

[8] Hsu MT, Coca-Prados M. Electron microscopic evidence for the circular form of RNA in the cytoplasm of eukaryotic cells. Nature. 1979;**280**:339-340. DOI: 10.1038/280339a0

[9] Cocquerelle C, Mascrez B, Hétuin D, Bailleul B. Mis-splicing yields circular RNA molecules. The FASEB Journal. 1993;**7**:155-160. DOI: 10.1096/fasebj.7.1.7678559

[10] Jeck WR, Sorrentino JA, Wang K, Slevin MK, Burd CE, Liu J, et al. Circular RNAs are abundant, conserved, and associated with ALU repeats. RNA. 2013;**19**:141-157. DOI: 10.1261/rna.035667.112

[11] Ashwal-Fluss R, Meyer M, Pamudurti NR, Ivanov A, Bartok O, Hanan M, et al. circRNA biogenesis competes with pre-mRNA splicing. Molecular Cell. 2014;**56**:55-66. DOI: 10.1016/j.molcel.2014.08.019

[12] Conn SJ, Pillman KA, Toubia J, Conn VM, Salmanidis M, Phillips CA, et al. The RNA binding protein quaking regulates formation of circRNAs. Cell. 2015;**160**:1125-1134. DOI: 10.1016/j.cell.2015.02.014

[13] Wang M, Yu F, Wu W, Zhang Y, Chang W, Ponnusamy M, et al. A novel type of non-coding RNA and their potential implications inantiviral immunity. International Journal of Biological Sciences. 2017;**13**:1497-1506. DOI: 10.7150/ijbs.22531

[14] Yu B, Shan G. Functions of long noncoding RNAs in the nucleus. Nucleus. 2016;7:155-166. DOI: 10.1080/19491034.2016.1179408

[15] Zhang Y, Zhang XO, Chen T, Xiang JF, Yin QF, Xing YH, et al. Circular intronic long noncoding RNAs. Molecular Cell. 2013;**51**:792-806. DOI: 10.1016/j.molcel.2013.08.017

[16] Li Y, Zheng Q, Bao C, Li S, Guo W, Zhao J, et al. Circular RNA is enriched and stable in exosomes: A promising biomarker for cancer diagnosis. Cell Research. 2015;**25**:981-984. DOI: 10.1038/cr.2015.82

[17] He J, Xie Q, Xu H, Li J, Li Y. Circular RNAs and cancer. Cancer Letters. 2017;**396**:138-144. DOI: 10.1016/j.canlet.2017.03.027

[18] Zhang Z, Yang T, Xiao J, Circular RNA. Promising biomarkers for human diseases. eBioMedicine. 2018;**34**: 267-274. DOI: 10.1016/j. ebiom.2018.07.036

[19] Yu CX, Sun S. An emerging role for circular RNAs in osteoarthritis. Yonsei Medical Journal. 2018;**59**:349-355. DOI: 10.3349/ymj.2018.59.3.349

[20] Zhang M, Xin Y. Circular RNAs: A new frontier for cancer diagnosis and therapy. Journal of Hematology & Oncology. 2018;**11**:21. DOI: 10.1186/ s13045-018-0569-5

[21] Kosik KS. Molecular biology: Circles reshape the RNA world. Nature. 2013; **495**:322-324. DOI: 10.1038/nature11956

[22] Liu L, Wang J, Khanabdali R, Kalionis B, Tai X, Xia S. Circular RNAs: Isolation, characterization and their potential role in diseases. RNA Biology. 2017;**14**:1715-1721. DOI: 10.1080/ 15476286.2017.1367886

[23] Yang L, Fu J, Zhou Y. Circular RNAs and their emerging roles in immune regulation. Frontiers in Immunology. 2018;**9**:2977. DOI: 10.3389/ fimmu.2018.02977

[24] Xu Z, Li P, Fan L, Wu M. The potential role of circRNA in tumor immunity regulation and immunotherapy. Frontiers in Immunology. 2018;**22**:9. DOI: 10.3389/ fimmu.2018.00009

[25] Wang M, Yu F, Li P. Circular RNAs: Characteristics, function and clinical significance in hepatocellular carcinoma. Cancers (Basel). 2018;**10**:pii: E258. DOI: 10.3390/cancers10080258

[26] Liu KS, Pan F, Mao XD, Liu C, Chen YJ. Biological functions of circular RNAs and their roles in occurrence of reproduction and gynecological diseases. American Journal of Translational Research. 2019;**11**: 1-15. Available from: https://www.ncbi. nlm.nih.gov/pmc/articles/PMC6 357300/

[27] Haque S, Harries LW. Circular RNAs (circRNAs) in health and disease. Genes (Basel). 2017;**8**:pii: E353. DOI: 10.3390/genes8120353

[28] Kristensen LS, Hansen TB, Venø MT, Kjems J. Circular RNAs in cancer: Opportunities and challenges in the field. Oncogene. 2018;**37**:555-565. DOI: 10.1038/onc.2017.361

[29] Wang Y, Lu T, Wang Q, Liu J, Jiao W. Circular RNAs: Crucial regulators in the human body (Review). Oncology Reports. 2018;**40**:3119-3135. DOI: 10.3892/or.2018.6733

[30] Hsiao KY, Sun HS, Tsai SJ. Circular RNA—New member of noncoding RNA with novel functions. Experimental Biology and Medicine (Maywood, N.J.). 2017;**242**:1136-1141. DOI: 10.1177/ 1535370217708978

[31] Hou LD, Zhang J. Circular RNAs: An emerging type of RNA in cancer. International Journal of Immunopathology and Pharmacology. 2017;**30**:1-6. DOI: 10.1177/ 0394632016686985

[32] Burd CE, Jeck WR, Liu Y, Sanoff HK, Wang Z, Sharpless NE. Expression of linear and novel circular forms of an INK4/ARFassociated non-coding RNA correlates with atherosclerosis risk. PLoS Genetics. 2010;**6**:e1001233. DOI: 10.1371/journal. pgen.1001233

[33] Meganck RM, Borchardt EK, Castellanos Rivera RM, Scalabrino ML, Wilusz JE, Marzluff WF, et al. Tissue-dependent expression and translation of circular RNAs with recombinant AAV vectors in vivo. Molecular Therapy:

Nucleic Acids. 2018;**13**:89-98. DOI: 10.1016/j.omtn.2018.08.008

[34] Liang WC, Wong CW, Liang PP, Shi M, Cao Y, Rao ST, et al. Translation of the circular RNA circβ-catenin promotes liver cancer cell growth through activation of the Wnt pathway. Genome Biology. 2019;**20**:84. DOI: 10.1186/s13059-019-1685-4

[35] Wang W, Wang Y, Piao H, Li B, Huang M, Zhu Z, et al. Circular RNAs as potential biomarkers and therapeutics for cardiovascular disease. PeerJ. 2019;**7**: e6831. DOI: 10.7717/peerj.6831

[36] Fan X, Weng X, Zhao Y, Chen W, Gan T, Xu D. Circular RNAs in cardiovascular disease: An overview. BioMed Research International. 2017; **2017**:5135781. DOI: 10.1155/2017/5135781

[37] Lee ECS, Elhassan SAM, Lim GPL, Kok WH, Tan SW, Leong EN, et al. The roles of circular RNAs in human development and diseases. Biomedicine & Pharmacotherapy. 2019; **111**:198-208. DOI: 10.1016/j.biopha.2018.12.052

[38] Li M, Ding W, Sun T, Tariq MA, Xu T, Li P, et al. Biogenesis of circular RNAs and their roles in cardiovascular development and pathology. The FEBS Journal. 2018;**285**:220-232. DOI: 10.1111/febs.14191

[39] Borghini A. Circular RNAs: Emerging players in coronary artery disease. Annals of Atherosclerosis Research. 2018;**1**:1-2. Available from: https://pdfs.semanticscholar.org/5d14/63fbe393f33601a5695ab0840b5738b6b8f3.pdf

[40] Carrara M, Fuschi P, Ivan C, Martelli F. Circular RNAs: Methodological challenges and perspectives in cardiovascular diseases. Journal of Cellular and Molecular Medicine. 2018;**22**:5176-5187. DOI: 10.1111/jcmm.13789

[41] Zhou MY, Yang JM, Xiong XD. The emerging landscape of circular RNA in cardiovascular diseases. Journal of Molecular and Cellular Cardiology. 2018;**122**:134-139. DOI: 10.1016/j.yjmcc.2018.08.012

[42] Zhao G. Significance of non-coding circular RNAs and micro RNAs in the pathogenesis of cardiovascular diseases. Journal of Medical Genetics. 2018;**55**: 713-720. DOI: 10.1136/jmedgenet-2018-105387

[43] Altesha MA, Ni T, Khan A, Liu K, Zheng X. Circular RNA in cardiovascular disease. Journal of Cellular Physiology. 2019;**234**: 5588-5600. DOI: 10.1002/jcp.27384

[44] Gong X, Wu G, Zeng C. Role of circular RNAs in cardiovascular diseases. Experimental Biology and Medicine (Maywood, N.J.). 2019;**244**: 73-82. DOI: 10.1177/1535370218822988

[45] Zeng Y, Du WW, Wu Y, Yang Z, Awan FM, Li X, et al. Circular RNA binds to and activates AKT phosphorylation and nuclear localization reducing apoptosis and enhancing cardiac repair. Theranostics. 2017;7:3842-3855. DOI: 10.7150/thno.19764

[46] Du WW, Yang W, Chen Y, Wu ZK, Foster FS, Yang Z, et al. Foxo3 circular RNA promotes cardiac senescence by modulating multiple factors associated with stress and senescence responses. European Heart Journal. 2017;**38**: 1402-1412. DOI: 10.1093/eurheartj/ehw001

[47] Holdt LM, Stahringer A, Sass K, Pichler G, Kulak NA, Wilfert W, et al. Circular non-coding RNA ANRIL modulates ribosomal RNA maturation and atherosclerosis in humans. Nature Communications. 2016;7:12429. DOI: 10.1038/ncomms12429

[48] Li CY, Ma L, Circular YB. RNA hsa_circ_0003575 regulates oxLDL induced

vascular endothelial cells proliferation and angiogenesis. Biomedicine & Pharmacotherapy. 2017;**95**:1514-1519. DOI: 10.1016/j.biopha.2017.09.064

[49] Dang RY, Liu FL, Li Y. Circular RNA hsa_circ_0010729 regulates vascular endothelial cell proliferation and apoptosis by targeting the miR-186/HIF-1alpha axis. Biochemical and Biophysical Research Communications. 2017;**490**:104-110. DOI: 10.1016/j.bbrc.2017.05.164

[50] Sun Y, Yang Z, Zheng B, Zhang XH, Zhang ML, Zhao XS, et al. A novel regulatory mechanism of smooth muscle alpha-actin expression by NRG-1/circACTA2/miR-548f-5p axis. Circulation Research. 2017;**121**:628-635. DOI: 10.1161/CIRCRESAHA.117.311441

[51] Chen J, Cui L, Yuan J, Zhang Y, Sang H. Circular RNA WDR77 target FGF-2 to regulate vascular smooth muscle cells proliferation and migration by sponging miR-124. Biochemical and Biophysical Research Communications. 2017;**494**:126-132. DOI: 10.1016/j.bbrc.2017.10.068

[52] Mao YY, Wang JQ, Guo XX, Bi Y, Wang CX. Circ-SATB2 upregulates STIM1 expression and regulates vascular smooth muscle cell proliferation and differentiation through miR-939. Biochemical and Biophysical Research Communications. 2018;**505**:119-125. DOI: 10.1016/j.bbrc.2018.09.069

[53] Memczak S, Jens M, Elefsinioti A, Torti F, Krueger J, Rybak A, et al. Circular RNAs are a large class of animal RNAs with regulatory potency. Nature. 2013;**495**:333-338. DOI: 10.1038/nature11928

[54] Bazan HA, Hatfield SA, Brug A, Brooks AJ, Lightell DJ Jr, Woods TC. Carotid plaque rupture is accompanied by an increase in the ratio of serum circR-284 to miR-221 levels. Circulation: Cardiovascular Genetics 2017;**10**:pii: e001720. DOI: 10.1161/CIRCGENETICS.117.001720

[55] Zhao Z, Li X, Gao C, Jian D, Hao P, Rao L, et al. Peripheral blood circular RNA hsa_circ_0124644 can be used as a diagnostic biomarker of coronary artery disease. Scientific Reports. 2017;7:39918. DOI: 10.1038/srep39918

[56] Wang L, Shen C, Wang Y, Zou T, Zhu H, Lu X, et al. Identification of circular RNA Hsa_circ_0001879 and Hsa_circ_0004104 as novel biomarkers for coronary artery disease. Atherosclerosis. 2019;**286**:88-96. DOI: 10.1016/j.atherosclerosis

[57] Geng HH, Li R, Su YM, Xiao J, Pan M, Cai XX, et al. The circular RNA Cdr1as promotes myocardial infarction by mediating the regulation of miR-7a on its target genes expression. PLoS One. 2016;**11**:e0151753. DOI: 10.1371/journal.pone.0151753

[58] Vausort M, Salgado-Somoza A, Zhang L, Leszek P, Scholz M, Teren A, et al. Myocardial infarction-associated circular RNA predicting left ventricular dysfunction. Journal of the American College of Cardiology. 2016;**68**:1247-1248. DOI: 10.1016/j.jacc.2016.06.040

[59] Salgado-Somoza A, Zhang L, Vausort M, Devaux Y. The circular RNA MICRA for risk stratification after myocardial infarction. International Journal of Cardiology. Heart & Vasculature. 2017;**17**:33-36. DOI: 10.1016/j.ijcha.2017.11.001

[60] Wang K, Gan TY, Li N, Liu CY, Zhou LY, Gao JN, et al. Circular RNA mediates cardiomyocyte death via miRNA-dependent upregulation of MTP18 expression. Cell Death and Differentiation. 2017;**24**:1111-1120. DOI: 10.1038/cdd.2017.61

[61] Wang K, Long B, Liu F, Wang JX, Liu CY, Zhao B, et al. A circular RNA

protects the heart from pathological hypertrophy and heart failure by targeting miR-223. European Heart Journal. 2016;**37**:2602-2611. DOI: 10.1093/eurheartj/ehv713

[62] Deng YY, Zhang W, She J, Zhang L, Chen T, Zhou J, et al. GW27-e1167 circular RNA related to PPAR[gamma] function as ceRNA of microRNA in human acute myocardial infarction. Journal of the American College of Cardiology. 2016;**68**:68-C52. DOI: 10.1016/j.jacc.2016.07.189

[63] Wu HJ, Zhang CY, Zhang S, Chang M, Wang HY. Microarray expression profile of circular RNAs in heart tissue of mice with myocardial infarction-induced heart failure. Cellular Physiology and Biochemistry. 2016;**39**:205-216. DOI: 10.1159/000445617

[64] Li M, Ding W, Tariq MA, Chang W, Zhang X, Xu W, et al. A circular transcript of ncx1 gene mediates ischemic myocardial injury by targeting miR-133a-3p. Theranostics. 2018;**8**:5855-5869. DOI: 10.7150/thno.27285

[65] Khan MA, Reckman YJ, Aufiero S, van den Hoogenhof MM, van der Made I, Beqqali A, et al. RBM20 regulates circular RNA production from the titin gene. Circulation Research. 2016;**119**:996-1003. DOI: 10.1161/CIRCRESAHA.116.309568

[66] Tan WL, Lim BT, Anene-Nzelu CG, Ackers-Johnson M, Dashi A, See K, et al. A landscape of circular RNA expression in the human heart. Cardiovascular Research. 2017;**113**:298-309. DOI: 10.1093/cvr/cvw250

[67] Liu C, Yao MD, Li CP, Shan K, Yang H, Wang JJ, et al. Silencing of circular RNA-ZNF609 ameliorates vascular endothelial dysfunction. Theranostics. 2017;**7**:2863-2877. DOI: 10.7150/thno.19353

[68] Wu N, Jin L, Cai J. Profiling and bioinformatics analyses reveal differential circular RNA expression in hypertensive patients. Clinical and Experimental Hypertension. 2017;**39**:454-459. DOI: 10.1080/10641963.2016.1273944

[69] Cheng X, Joe B. Circular RNAs in rat models of cardiovascular and renal diseases. Physiological Genomics. 2017;**49**:484-490. DOI: 10.1152/physiolgenomics.00064.2017

[70] Zheng S, Gu T, Bao X, Sun J, Zhao J, Zhang T, et al. Circular RNA hsa_circ_0014243 may serve as a diagnostic biomarker for essential hypertension. Experimental and Therapeutic Medicine. 2019;**17**:1728-1736. DOI: 10.3892/etm.2018.7107

[71] Bao X, Zheng S, Mao S, Gu T, Liu S, Sun J, et al. A potential risk factor of essential hypertension in case-control study: Circular RNA hsa_circ_0037911. Biochemical and Biophysical Research Communications. 2018;**498**:789-794. DOI: 10.1016/j.bbrc.2018.03.059

[72] Liu L, Gu T, Bao X, Zheng S, Zhao J, Zhang L. Microarray profiling of circular RNA identifies hsa_circ_0126991 as a potential risk factor for essential hypertension. Cytogenetic and Genome Research. 2019;**157**:203-212. DOI: 10.1159/000500063

[73] Westholm JO, Miura P, Olson S, Shenker S, Joseph B, Sanfilippo P, et al. Genome-wide analysis of drosophila circular RNAs reveals their structural and sequence properties and age-dependent neural accumulation. Cell Reports. 2014;**9**:1966-1980. DOI: 10.1016/j.celrep.2014.10.062

[74] Rybak-Wolf A, Stottmeister C, Glažar P, Jens M, Pino N, Giusti S, et al. Circular RNAs in the mammalian brain are highly abundant, conserved, and dynamically expressed. Molecular Cell.

2015;**58**:870-885. DOI: 10.1016/j.molcel.2015.03.027

[75] You X, Vlatkovic I, Babic A, Will T, Epstein I, Tushev G, et al. Neural circular RNAs are derived from synaptic genes and regulated by development and plasticity. Nature Neuroscience. 2015;**18**:603-610. DOI: 10.1038/nn.3975

[76] Chen W, Schuman E. Circular RNAs in brain and other tissues: A functional enigma. Trends in Neurosciences. 2016;**39**:597-604. DOI: 10.1016/j.tins.2016.06.006

[77] Zhang Y, Xue W, Li X, Zhang J, Chen S, Zhang JL, et al. The biogenesis of nascent circular RNAs. Cell Reports. 2016;**15**:611-624. DOI: 10.1016/j.celrep.2016.03.058

[78] Piwecka M, Glažar P, Hernandez-Miranda LR, Memczak S, Wolf SA, Rybak-Wolf A, et al. Loss of a mammalian circular RNA locus causes miRNA deregulation and affects brain function. Science. 2017;**357**:pii: eaam8526. DOI: 10.1126/science.aam8526

[79] Enuka Y, Lauriola M, Feldman ME, Sas-Chen A, Ulitsky I, Yarden Y. Circular RNAs are long-lived and display only minimal early alterations in response to a growth factors. Nucleic Acids Research. 2016;**44**:1370-1383. DOI: 10.1093/nar/gkv1367

[80] Maiese K. Disease onset and aging in the world of circular RNAs. Journal of Translational Science. 2016;**2**:327-329. Available from: https://www.ncbi.nlm.nih.gov/pmc/articles/PMC5026119/

[81] Panda AC, Grammatikakis I, Kim KM, De S, Martindale JL, Munk R, et al. Identification of senescence-associated circular RNAs (SAC-RNAs) reveals senescence suppressor CircPVT1. Nucleic Acids Research. 2017;**45**:4021-4035. DOI: 10.1093/nar/gkw1201

[82] Knupp D, Miura P. CircRNA accumulation: A new hallmark of aging? Mechanisms of Ageing and Development. 2018;**173**:71-79. DOI: 10.1016/j.mad.2018.05.001

[83] Lukiw WJ, Circular RNA. (circRNA) in Alzheimer's disease (AD). Frontiers in Genetics. 2013;**4**:307. DOI: 10.3389/fgene.2013.00307

[84] Bingol B, Sheng M. Deconstruction for reconstruction: The role of proteolysis in neural plasticity and disease. Neuron. 2011;**69**:22-32. DOI: 10.1016/j.neuron.2010.11.006

[85] Hansen TB, Jensen TI, Clausen BH, Bramsen JB, Finsen B, Damgaard CK, et al. Natural RNA circles function as efficient microRNA sponges. Nature. 2013;**495**(7441):384-388. DOI: 10.1038/nature11993

[86] Wang X, Tan L, Lu Y, Peng J, Zhu Y, Zhang Y, et al. MicroRNA-138 promotes tau phosphorylation by targeting retinoic acid receptor alpha. FEBS Letters. 2015;**589**:726-729. DOI: 10.1016/j.febslet.2015.02.001

[87] Tatro ET, Risbrough V, Soontornniyomkij B, Young J, Shumaker-Armstrong S, Jeste DV, et al. Short-term recognition memory correlates with regional CNS expression of microRNA-138 in mice. The American Journal of Geriatric Psychiatry. 2013;**21**:461-473. DOI: 10.1016/j.jagp.2012.09.005

[88] Junn E, Lee KW, Jeong BS, Chan TW, Im JY, Mouradian MM. Repression of alpha-synuclein expression and toxicity by microRNA-7. Proceedings of the National Academy of Sciences of the United States of America. 2009;**106**:13052-13057. DOI: 10.1073/pnas.0906277106

[89] Choi SY, Pang K, Kim JY, Ryu JR, Kang H, Liu Z, et al. Post-transcriptional regulation of SHANK3 expression by

microRNAs related to multiple neuropsychiatric disorders. Molecular Brain. 2015;**8**:74. DOI: 10.1186/s13041-015-0165-3

[90] Zhang J, Sun XY, Zhang LY. MicroRNA-7/Shank3 axis involved in schizophrenia pathogenesis. Journal of Clinical Neuroscience. 2015;**22**:1254-1257. DOI: 10.1016/j.jocn.2015.01.031

[91] Satoh J, Yamamura T. Gene expression profile following stable expression of the cellular prion protein. Cellular and Molecular Neurobiology. 2004;**24**:793-814. DOI: 10.1007/s10571-004-6920-0

[92] Satoh J, Obayashi S, Misawa T, Sumiyoshi K, Oosumi K, Tabunoki H. Protein microarray analysis identifies human cellular prion protein interactors. Neuropathology and Applied Neurobiology. 2009;**35**:16-35. DOI: 10.1111/j.1365-2990.2008.00947.x

[93] Wang YH, Yu XH, Luo SS, Han H. Comprehensive circular RNA profiling reveals that circular RNA100783 is involved in chronic CD28-associated CD8(+)T cell ageing. Immunity & Ageing. 2015;**12**:17. DOI: 10.1186/s12979-015-0042-z

[94] Fu D, Yu W, Li M, Wang H, Liu D, Song X, et al. MicroRNA-138 regulates the balance of Th1/Th2 via targeting RUNX3 in psoriasis. Immunology Letters. 2015;**166**:55-62. DOI: 10.1016/j.imlet.2015.05.014

[95] Xu H, Guo S, Li W, Yu P. The circular RNA Cdr1as, via miR-7 and its targets, regulates insülin transcription and secretion in islet cells. Scientific Reports. 2015;**5**:12453. DOI: 10.1038/srep12453

[96] Stoll L, Sobel J, Rodriguez-Trejo A, Guay C, Lee K, Venø MT, et al. Circular RNAs as novel regulators of β-cell functions in normal and disease conditions. Molecular Metabolism. 2018;**9**:69-83. DOI: 10.1016/j.molmet.2018.01.010

[97] Li P, Yang X, Yuan W, Yang C, Zhang X, Han J, et al. CircRNA-Cdr1as exerts anti-oncogenic functions in bladder cancer by sponging MicroRNA-135a. Cellular Physiology and Biochemistry. 2018;**46**:1606-1616. DOI: 10.1159/000489208

[98] Ledford H. Circular RNAs throw genetics for a loop. Nature. 2013;**494**:415. DOI: 10.1038/494415a

[99] Lonskaya I, Shekoyan AR, Hebron ML, Desforges N, Algarzae NK, Moussa CE. Diminished parkin solubility and colocalization with intraneuronal amyloid-beta are associated with autophagic defects in Alzheimer's disease. Journal of Alzheimer's Disease. 2013;**33**:231-247. DOI: 10.3233/JAD-2012-121141

[100] Zhao Y, Alexandrov PN, Jaber V, Lukiw WJ. Deficiency in the ubiquitin conjugating enzyme UBE2A in Alzheimer's Disease (AD) is linked to deficits in a natural circular miRNA-7 sponge (circRNA; ciRS-7). Genes (Basel). 2016;**7**:116. DOI: 10.3390/genes7120116

[101] Shi Z, Chen T, Yao Q, Zheng L, Zhang Z, Wang J, et al. The circular RNA ciRS-7 promotes APP and BACE1 degradation in an NF-κB-dependent manner. The FEBS Journal. 2017;**284**:1096-1109. DOI: 10.1111/febs.14045

[102] Schröder J, Ansaloni S, Schilling M, Liu T, Radke J, Jaedicke M, et al. MicroRNA-138 is a potential regulator of memory performance in humans. Frontiers in Human Neuroscience. 2014;**8**:501. DOI: 10.3389/fnhum.2014.00501

[103] Kalia LV, Lang AE. Parkinson's disease. Lancet. 2015;**386**:896-912. DOI: 10.1016/S0140-6736(14)61393-3

[104] Schapira AH, Jenner P. Etiology and pathogenesis of Parkinson's disease. Movement Disorders. 2011;**26**: 1049-1055. DOI: 10.1002/mds.23732

[105] Thomas B, Beal MF. Parkinson's disease. Human Molecular Genetics. 2007;**16**:R183-R194. DOI: 10.1093/hmg/ddm159

[106] Lim KL, Dawson VL, Dawson TM. The cast of molecular characters in Parkinson's disease: Felons, conspirators, and suspects. Annals of the New York Academy of Sciences. 2003;**991**:80-92. DOI: 10.1111/j.1749-6632.2003.tb07465.x

[107] Baba M, Nakajo S, Tu PH, Tomita T, Nakaya K, Lee VM, et al. Aggregation of alphasynuclein in Lewy bodies of sporadic Parkinson's disease and dementia with Lewy bodies. The American Journal of Pathology. 1998; **152**:879-884. Available from: https://www.ncbi.nlm.nih.gov/pmc/articles/PMC1858234/

[108] Rodriguez JA, Ivanova MI, Sawaya MR, Cascio D, Reyes FE, Shi D, et al. Structure of the toxic core of α-synuclein from invisible crystals. Nature. 2015;**525**:486-490. DOI: 10.1038/nature15368

[109] Hancock JM. Circles within circles: Commentary on Ghosal et al. (2013) Circ2Traits: A comprehensive database for circular RNA potentially associated with disease and traits. Frontiers in Genetics. 2015;**5**:459. DOI: 10.3389/fgene.2014.00459

[110] Ghosal S, Das S, Sen R, Basak P, Chakrabarti J. Circ2Traits: A comprehensive database for circular RNA potentially associated with disease and traits. Frontiers in Genetics. 2013;**4**: 283. DOI: 10.3389/fgene.2013.00283

[111] Sang Q, Liu X, Wang L, Qi L, Sun W, Wang W, et al. CircSNCA downregulation by pramipexole treatment mediates cell apoptosis and autophagy in Parkinson's disease by targeting miR-7. Aging (Albany NY). 2018;**10**:1281-1293. DOI: 10.18632/aging.101466

[112] Ghavami S, Shojaei S, Yeganeh B, Ande SR, Jangamreddy JR, Mehrpour M, et al. Autophagy and apoptosis dysfunction in neurodegenerative disorders. Progress in Neurobiology. 2014;**112**:24-49. DOI: 10.1016/j.pneurobio.2013.10.004

[113] Kountouras J, Zavos C, Polyzos SA, Deretzi G, Vardaka E, Giartza-Taxidou E, et al. Helicobacter pylori infection and Parkinson's disease: Apoptosis as an underlying common contributor. European Journal of Neurology. 2012; **19**:e56. DOI: 10.1111/j.1468-1331.2012.03695.x

[114] Chen X, Yang T, Wang W, Xi W, Zhang T, Li Q, et al. Circular RNAs in immune responses and immune diseases. Theranostics. 2019;**9**:588-607. DOI: 10.7150/thno.29678

[115] Zhou Z, Sun B, Huang S, Zhao L. Roles of circular RNAs in immune regulation and autoimmune diseases. Cell Death & Disease. 2019;**10**:503. DOI: 10.1038/s41419-019-1744-5

[116] Xia X, Tang X, Wang S. Roles of CircRNAs in autoimmune diseases. Frontiers in Immunology. 2019;**10**:639. DOI: 10.3389/fimmu.2019.00639

[117] Li H, Li K, Lai W, Li X, Wang H, Yang J, et al. Comprehensive circular RNA profiles in plasma reveals that circular RNAs can be used as novel biomarkers for systemic lupus erythematosus. Clinica Chimica Acta. 2018;**480**:17-25. DOI: 10.1016/j.cca.2018.01.026

[118] Wu XN, Ye YX, Niu JW, Li Y, Li X, You X, et al. Defective PTEN regulation

contributes to B cell hyperresponsiveness in systemic lupus erythematosus. Science Translational Medicine. 2014;**6**:246ra99. DOI: 10.1126/scitranslmed.3009131

[119] Ouyang Q, Wu J, Jiang Z, Zhao J, Wang R, Lou A, et al. Microarray expression profile of circular RNAs in peripheral blood mononuclear cells from rheumatoid arthritis patients. Cellular Physiology and Biochemistry. 2017;**42**:651-659. DOI: 10.1159/000477883

[120] Wang W, Zhang Y, Zhu B, Duan T, Xu Q, Wang R, et al. Plasma microRNA expression profiles in Chinese patients with rheumatoid arthritis. Oncotarget. 2015;**6**:42557-42568. DOI: 10.18632/oncotarget.6449

[121] Zheng F, Yu X, Huang J, Dai Y. Circular RNA expression profiles of peripheral blood mononuclear cells in rheumatoid arthritis patients, based on microarray chip technology. Molecular Medicine Reports. 2017;**16**:8029-8036. DOI: 10.3892/mmr.2017.7638

[122] Murata K, Yoshitomi H, Tanida S, Ishikawa M, Nishitani K, Ito H, et al. Plasma and synovial fluid microRNAs as potential biomarkers of rheumatoid arthritis and osteoarthritis. Arthritis Research & Therapy. 2010;**12**:R86. DOI: 10.1186/ar3013

[123] Xu K, Xu P, Yao JF, Zhang YG, Hou WK, Lu SM. Reduced apoptosis correlates with enhanced autophagy in synovial tissues of rheumatoid arthritis. Inflammation Research. 2013;**62**: 229-237. DOI: 10.1007/s00011-012-0572-1

[124] Iparraguirre L, Muñoz-Culla M, Prada-Luengo I, Castillo-Triviño T, Olascoaga J, Otaegui D. Circular RNA profiling reveals that circular RNAs from ANXA2 can be used as new biomarkers for multiple sclerosis. Human Molecular Genetics. 2017;**26**: 3564-3572. DOI: 10.1093/hmg/ddx243

[125] Paraboschi EM, Rimoldi V, Soldà G, Tabaglio T, Dall'Osso C, Saba E, et al. Functional variations modulating PRKCA expression and alternative splicing predispose to multiple sclerosis. Human Molecular Genetics. 2014;**23**:6746-6761. DOI: 10.1093/hmg/ddu392

[126] Kang GJ, Lee HJ, Byun HJ, Kim EJ, Kim HJ, Park MK, et al. Novel involvement of miR-522-3p in high-mobility group box 1-induced prostaglandin reductase 1 expression and reduction of phagocytosis. Biochimica et Biophysica Acta—Molecular Cell Research. 2017;**1864**: 625-633. DOI: 10.1016/j.bbamcr.2017.01.006

[127] Danza K, De Summa S, Pinto R, Pilato B, Palumbo O, Carella M, et al. miRNA regulation in familial and sporadic breast cancer. Oncotarget. 2017;**8**:50715-50723. DOI: 10.18632/oncotarget.14899

[128] Maass PG, Glažar P, Memczak S, Dittmar G, Hollfinger I, Schreyer L, et al. A map of human circular RNAs in clinically relevant tissues. Journal of Molecular Medicine (Berlin, Germany). 2017;**95**:1179-1189. DOI: 10.1007/s00109-017-1582-9

[129] Brigida I, Sauer AV, Ferrua F, Giannelli S, Scaramuzza S, Pistoia V, et al. B-cell development and functions and therapeutic options in adenosine deaminase-deficient patients. Journal of Allergy and Clinical Immunology. 2014; **133**:799-806.e10. DOI: 10.1016/j.jaci.2013.12.1043

[130] Cassani B, Mirolo M, Cattaneo F, Benninghoff U, Hershfield M, Carlucci F, et al. Altered intracellular and extracellular signaling leads to impaired T-cell functions in ADA-SCID

patients. Blood. 2008;**111**:4209-4219. DOI: 10.1182/blood-2007-05-092429

[131] Sheldon H, Andre M, Legg JA, Heal P, Herbert JM, Sainson R, et al. Active involvement of Robo1 and Robo4 in filopodia formation and endothelial cell motility mediated via WASP and other actin nucleation-promoting factors. The FASEB Journal. 2009;**23**: 513-522. DOI: 10.1096/fj.07-098269

[132] Stewart B, Wild CP. International Agency for Research on Cancer. Lyon, France: WHO: World Health Organization; 2017. World cancer report 2014. Available from: https://www.who.int/cancer/publications/WRC_2014/en/

[133] Miller KD, Nogueira L, Mariotto AB, Rowland JH, Yabroff KR, Alfano CM, et al. Cancer treatment and survivorship statistics, 2019. CA: A Cancer Journal for Clinicians. 2019;**69**: 363-385. DOI: 10.3322/caac.21565

[134] Ferlay J, Soerjomataram I, Dikshit R, Eser S, Mathers C, Rebelo M, et al. Cancer incidence and mortality worldwide: Sources, methods and major patterns in GLOBOCAN 2012. International Journal of Cancer. 2015; **136**:E359-E386. DOI: 10.1002/ijc.29210

[135] Osada H, Takahashi T. MicroRNAs in biological processes and carcinogenesis. Carcinogenesis. 2007;**28**: 2-12. DOI: 10.1093/carcin/bgl185

[136] Flynt AS, Lai EC. Biological principles of microRNA-mediated regulation: Shared themes amid diversity. Nature Reviews. Genetics. 2008;**9**:831-842. DOI: 10.1038/nrg2455

[137] Chen CZ. MicroRNAs as oncogenes and tumor suppressors. The New England Journal of Medicine. 2005;**353**: 1768-1771. DOI: 10.1056/NEJMp058190

[138] Huang MS, Liu JY, Xia XB, Liu YZ, Li X, Yin JY, et al. Hsa_circ_0001946 inhibits lung cancer progression and mediates cisplatin sensitivity in non-small cell lung cancer via the nucleotide excision repair signaling pathway. Frontiers in Oncology. 2019;**9**:508. DOI: 10.3389/fonc.2019.00508

[139] Yao Y, Hua Q, Zhou Y, Shen H. CircRNA has_circ_0001946 promotes cell growth in lung adenocarcinoma by regulating miR-135a-5p/SIRT1 axis and activating Wnt/β-catenin signaling pathway. Biomedicine & Pharmacotherapy. 2019;**111**:1367-1375. DOI: 10.1016/j.biopha.2018.12.120

[140] Xue YB, Ding MQ, Xue L, Luo JH. CircAGFG1 sponges miR-203 to promote EMT and metastasis of non-small-cell lung cancer by upregulating ZNF281 expression. Thoracic Cancer. 2019;**10**:1692-1701. DOI: 10.1111/1759-7714.13131

[141] Wei S, Zheng Y, Jiang Y, Li X, Geng J, Shen Y, et al. The circRNA circPTPRA suppresses epithelial-mesenchymal transitioning and metastasis of NSCLC cells by sponging miR-96-5p. eBioMedicine. 2019;**44**: 182-193. DOI: 10.1016/j.ebiom.2019.05.032

[142] Wan J, Hao L, Zheng X, Li Z. Circular RNA circ_0020123 promotes non-small cell lung cancer progression by acting as a ceRNA for miR-488-3p to regulate ADAM9expression. Biochemical and Biophysical Research Communications. 2019;**515**(2):303-309. DOI: 10.1016/j.bbrc.2019.05.158

[143] Qu D, Yan B, Xin R, Ma T. A novel circular RNA hsa_circ_0020123 exerts oncogenic properties through suppression of miR-144 in non-small cell lung cancer. American Journal of Cancer Research. 2018;**8**:1387-1402. Available from: https://www.ncbi.nlm.nih.gov/pmc/articles/PMC6129481/

[144] Wang L, Ma H, Kong W, Liu B, Zhang X. Up-regulated circular RNA VANGL1 contributes to progression of non-small cell lung cancer through inhibition ofmiR-195 and activation of

Bcl-2. Bioscience Reports. 2019;**39**:pii: BSR20182433. DOI: 10.1042/BSR20182433

[145] Wang T, Wang X, Du Q, Wu N, Liu X, Chen Y, et al. The circRNA circP4HB promotes NSCLC aggressiveness and metastasis by sponging miR-133a-5p. Biochemical and Biophysical Research Communications. 2019;**513**:904-911. DOI: 10.1016/j.bbrc.2019.04.108

[146] Chang H, Qu J, Wang J, Liang X, Sun W. Circular RNA circ_0026134 regulates non-small cell lung cancer cell proliferation and invasion via sponging miR-1256 and miR-1287. Biomedicine & Pharmacotherapy. 2019;**112**:108743. DOI: 10.1016/j.biopha.2019.108743

[147] Liu G, Shi H, Deng L, Zheng H, Kong W, Wen X, et al. Circular RNA circ-FOXM1 facilitates cell progression as ceRNA to target PPDPF and MACC1 by sponging miR-1304-5p in non-small cell lung cancer. Biochemical and Biophysical Research Communications. 2019;**513**:207-212. DOI: 10.1016/j.bbrc.2019.03.213

[148] An J, Shi H, Zhang N, Song S. Elevation of circular RNA circ_0003645 forecasts unfavorable prognosis and facilitates cell progression via miR-1179/TMEM14A pathway in non-small cell lung cancer. Biochemical and Biophysical Research Communications. 2019;**511**:921-925. DOI: 10.1016/j.bbrc.2019.03.011

[149] Yan Y, Zhang R, Zhang X, Zhang A, Zhang Y, Bu X. RNA-Seq profiling of circular RNAs and potential function of hsa_circ_0002360 in human lung adenocarcinom. American Journal of Translational Research. 2019;**11**: 160-175. Available from: https://www.ncbi.nlm.nih.gov/pmc/articles/PMC6357307/

[150] Chen L, Nan A, Zhang N, Jia Y, Li X, Ling Y, et al. Circular RNA 100146 functions as an oncogene through direct binding to miR-361-3p and miR-615-5p in non-small cell lung cancer. Molecular Cancer. 2019;**18**:13. DOI: 10.1186/s12943-019-0943-0

[151] Qiu BQ, Zhang PF, Xiong D, Xu JJ, Long X, Zhu SQ, et al. CircRNA fibroblast growth factor receptor 3 promotes tumor progression in non-small cell lung cancer by regulating Galectin-1-AKT/ERK1/2 signaling. Journal of Cellular Physiology. 2019; **234**:11256-11264. DOI: 10.1002/jcp.27783

[152] Yao Y, Hua Q, Zhou Y. CircRNA has_circ_0006427 suppresses the progression of lung adenocarcinoma by regulating miR-6783-3p/DKK1 axis and inactivating Wnt/β-catenin signaling pathway. Biochemical and Biophysical Research Communications. 2019;**508**:37-45. DOI: 10.1016/j.bbrc.2018.11.079

[153] Wang L, Tong X, Zhou Z, Wang S, Lei Z, Zhang T, et al. Circular RNA hsa_circ_0008305 (circPTK2) inhibits TGF-β-induced epithelial-mesenchymal transition and metastasis by controlling TIF1γ in non-small cell lung cancer. Molecular Cancer. 2018;**17**:140. DOI: 10.1186/s12943-018-0889-7

[154] Chen D, Ma W, Ke Z, Xie F. CircRNA hsa_circ_100395 regulates miR-1228/TCF21 pathway to inhibit lung cancer progression. Cell Cycle. 2018;**17**:2080-2090. DOI: 10.1080/15384101.2018.1515553

[155] Han J, Zhao G, Ma X, Dong Q, Zhang H, Wang Y, et al. CircRNA circ-BANP-mediated miR-503/LARP1 signaling contributes to lung cancer progression. Biochemical and Biophysical Research Communications. 2018;**503**:2429-2435. DOI: 10.1016/j.bbrc.2018.06.172

[156] Ma X, Yang X, Bao W, Li S, Liang S, Sun Y, et al. circMAN2B2 facilitates lung cancer cell proliferation and invasion via miR-1275/FOXK1 axis.

Biochemical and Biophysical Research Communications. 2018;**498**:1009-1015. DOI: 10.1016/j.bbrc.2018.03.105

[157] Li Y, Hu J, Li L, Cai S, Zhang H, Zhu X, et al. Upregulated circular RNA circ_0016760 indicates unfavorable prognosis in NSCLC and promotes cell progression through miR-1287/GAGE1 axis. Biochemical and Biophysical Research Communications. 2018;**503**: 2089-2094. DOI: 10.1016/j.bbrc.2018.07.164

[158] Gao D, Zhang X, Liu B, Meng D, Fang K, Guo Z, et al. Screening circular RNA related to chemotherapeutic resistance in breast cancer. Epigenomics. 2017;**9**:1175-1188. DOI: 10.2217/epi-2017-0055

[159] Tang H, Huang X, Wang J, Yang L, Kong Y, Gao G, et al. circKIF4A acts as a prognostic factor and mediator to regulate the progression of triple-negative breast cancer. Molecular Cancer. 2019;**18**:23. DOI: 10.1186/s12943-019-0946-x

[160] Xie R, Tang J, Zhu X, Jiang H. Silencing of hsa_circ_0004771 inhibits proliferation and induces apoptosis in breast cancer through activation of miR-653 by targeting ZEB2 signaling pathway. Bioscience Reports. 2019;**39**. DOI: 10.1042/BSR20181919

[161] Xu JZ, Shao CC, Wang XJ, Zhao X, Chen JQ, Ouyang YX, et al. circTADA2As suppress breast cancer progression and metastasis via targeting miR-203a-3p/SOCS3 axis. Cell Death & Disease. 2019;**10**:175. DOI: 10.1038/s41419-019-1382-y

[162] Yang R, Xing L, Zheng X, Sun Y, Wang X, Chen J. The circRNA circAGFG1 acts as a sponge of miR-195-5p to promote triple-negative breast cancer progression through regulating CCNE1 expression. Molecular Cancer. 2019;**18**:4. DOI: 10.1186/s12943-018-0933-7

[163] Chen B, Wei W, Huang X, Xie X, Kong Y, Dai D, et al. circEPSTI1 as a prognostic marker and mediator of triple-negative breast cancer progression. Theranostics. 2018;**8**: 4003-4015. DOI: 10.7150/thno.24106

[164] Liu Y, Lu C, Zhou Y, Zhang Z, Sun L. Circular RNA hsa_circ_0008039 promotes breast cancer cell proliferation and migration by regulating miR-432-5p/E2F3 axis. Biochemical and Biophysical Research Communications. 2018;**502**:358-363. DOI: 10.1016/j.bbrc.2018.05.166

[165] Song L, Xiao Y. Downregulation of hsa_circ_0007534 suppresses breast cancer cell proliferation and invasion by targeting miR-593/MUC19 signal pathway. Biochemical and Biophysical Research Communications. 2018;**503**: 2603-2610. DOI: 10.1016/j.bbrc.2018.08.007

[166] Wang H, Xiao Y, Wu L, Ma D. Comprehensive circular RNA profiling reveals the regulatory role of the circRNA-000911/miR-449a pathway in breast carcinogenesis. International Journal of Oncology. 2018;**52**:743-754. DOI: 10.3892/ijo.2018.4265

[167] Wang S, Li Q, Wang Y, Li X, Wang R, Kang Y, et al. Upregulation of circ-UBAP2 predicts poor prognosis and promotes triple-negative breast cancer progression through the miR-661/MTA1 pathway. Biochemical and Biophysical Research Communications. 2018;**505**: 996-1002. DOI: 10.1016/j.bbrc.2018.10.026

[168] Wu J, Jiang Z, Chen C, Hu Q, Fu Z, Chen J, et al. CircIRAK3 sponges miR-3607 to facilitate breast cancer metastasis. Cancer Letters. 2018;**430**: 179-192. DOI: 10.1016/j.canlet.2018.05.033

[169] Xu Y, Yao Y, Leng K, Ji D, Qu L, Liu Y, et al. Increased expression of circular RNA circ_0005230 indicates

dismal prognosis in breast cancer and regulates cell proliferation and invasion via miR-618/CBX8 signal pathway. Cellular Physiology and Biochemistry. 2018;**51**:1710-1722. DOI: 10.1159/000495675

[170] Zeng K, He B, Yang BB, Xu T, Chen X, Xu M, et al. The pro-metastasis effect of circANKS1B in breast cancer. Molecular Cancer. 2018;**17**:160. DOI: 10.1186/s12943-018-0914-x

[171] He R, Liu P, Xie X, Zhou Y, Liao Q, Xiong W, et al. circGFRA1 and GFRA1 act as ceRNAs in triple negative breast cancer by regulating miR-34a. Journal of Experimental & Clinical Cancer Research. 2017;**36**:145. DOI: 10.1186/s13046-017-0614-1

[172] Chen X, Han P, Zhou T, Guo X, Song X, Li Y. circRNADb: A comprehensive database for human circular RNAs with protein-coding annotations. Scientific Reports. 2016;**6**: 34985. DOI: 10.1038/srep34985

[173] Xu S, Zhou L, Ponnusamy M, Zhang L, Dong Y, Zhang Y, et al. A comprehensive review of circRNA: From purification and identification to disease marker potential. PeerJ. 2018;**6**: e5503. DOI: 10.7717/peerj.5503

[174] Li JH, Liu S, Zhou H, Qu LH, Yang JH. starBase v2.0: Decoding miRNA-ceRNA, miRNA-ncRNA and protein-RNA interaction networks from large-scale CLIP-Seq data. Nucleic Acids Research. 2014;**42**:D92-D97. DOI: 10.1093/nar/gkt1248

[175] Dudekula DB, Panda AC, Grammatikakis I, De S, Abdelmohsen K, Gorospe M. CircInteractome: A web tool for exploring circular RNAs and their interacting proteins and microRNAs. RNA Biology. 2016;**13**:34-42. DOI: 10.1080/15476286.2015.1128065

[176] Xia S, Feng J, Lei L, Hu J, Xia L, Wang J, et al. Comprehensive characterization of tissue-specific circular RNAs in the human and Mouse genomes. Briefings in Bioinformatics. 2016;**18**:984-992. DOI: 10.1093/bib/bbw081

[177] Zhang XO, Dong R, Zhang Y, Zhang JL, Luo Z, Zhang J, et al. Diverse alternative back-splicing and alternative splicing landscape of circular RNAs. Genome Research. 2016;**26**:1277-1287. DOI: 10.1101/gr.202895.115

[178] Liu M, Wang Q, Shen J, Yang BB, Ding X. Circbank: A comprehensive database for circRNA with standard nomenclature. RNA Biology. 2019;**16**: 899-905. DOI: 10.1080/15476286.2019.1600395

Chapter 7

Evaluation of the Synergistic Effect of Amikacin with Cefotaxime against *Pseudomonas aeruginosa* and Its Biofilm Genes Expression

Azza S. El-Demerdash and Neveen R. Bakry

Abstract

A total of 100 broiler chickens were examined for the presence of *Pseudomonas aeruginosa* by standard microbiological techniques. Susceptibility pattern for amikacin and cefotaxime was performed by Kirby-Bauer and microdilution assays. Then, checkerboard titration in trays was applied and FIC was measured to identify the type of interaction between the two antibiotics. The ability of isolates to form in vitro biofilm was detected by two methods, one qualitative (CRA) and the other quantitative (MTP), followed by investigating the effect of each antibiotic alone and in combination on the expression of biofilm genes. The overall isolation percentage of *P. aeruginosa* was 21%. Resistance to each antibiotic was more than 50%; the range of cefotaxime MIC was 8–512 μg/ml, while amikacin MIC range was 1–64 μg/ml. The FIC index established a synergistic association between tested two drugs in 17 (81%) of isolates and the remaining represent partially synergism. The qualitative technique showed that only 66.6% of the isolates were considered biofilm producers, while the quantitative technique showed that 90.4% of the isolates were biofilm producers. Further to RT-PCR investigation, significant repression against biofilm-forming genes (*filC*, *pelA*, and *pslA*) was observed for the combination of antibiotics when compared with monotherapy.

Keywords: *P. aeruginosa*, cefotaxime, amikacin, combination therapy, biofilm, gene expression

1. Introduction

The infection with *Pseudomonas aeruginosa* is responsible for humanity in poultry and clinical signs including respiratory signs and septicaemia. *P. aeruginosa* produces dyspnea and cheesy deposits on the serous surfaces lining the air sacs and peritoneal cavity and also congestion of the internal organs, perihepatitis, and pericarditis [1]. *Pseudomonas* species are not related to disease entity except *Pseudomonas aeruginosa* that has been associated with infection in both man and animals. The disease of pseudomonas induces a significant economic loss to the farm by causing high mortality of newly hatched chickens and death of embryo at a later stage [2]. Furthermore, *Pseudomonas aeruginosa* shows innate resistance to almost antibiotics in recent years [3, 4].

Due to this intrinsic resistance to antibiotics, its ability to easily develop new resistance, its ability to create biofilms, and the recent decline in drug discovery programs, *P. aeruginosa* infections have become an urgent worldwide health concern [3, 5]. Recent efforts to focus on this rising challenge comprise repositioning screens to recognize commercially permitted drugs with novel antimicrobial activity [6–9] and combinatorial drug screens to categorize combinations of traditional antibiotics and novel repositionable modulators [10, 11].

Concomitant use of antibiotics (combination therapy) is recommended for severe infections when *P. aeruginosa* is the suspected pathogen, to prevent the development of resistance during treatment and to achieve a wide spectrum of activity. In addition to preventing the development of resistance, the combined use of antibiotics (as cephalosporins and aminoglycosides) may have synergistic effects and may reduce the occurrence of side effects, since each drug is used at a lower dose than would be used for monotherapy [12].

Concerning bacterial biofilms, Batoni et al. [13] and Grassi et al. [14] proved a strong interaction between the effectiveness of combination therapy and biofilms formed by *P. aeruginosa*. Therefore, the present study concerned the effect of cefotaxime, amikacin singly, and in combination besides validating the activity of them on biofilm expression of the obtained *P. aeruginosa* isolates.

2. Material and methods

2.1 Sampling and isolate characterization

A total of 500 samples of the liver, heart, kidney, spleen, and lung (100 each) was aseptically collected from 100 freshly dead and diseased with respiratory manifestations broiler chickens from different ages and localities in Sharkia province, Egypt, from November 2018 to February 2019. All samples were subjected to conventional methods for isolation and identification of pseudomonas recommended by the Health Protection Agency [15]. *Pseudomonas aeruginosa* isolates were further identified with API20E kits (BioMérieux, France).

2.2 Antibiotic susceptibility testing

2.2.1 Disk diffusion method

The antimicrobial susceptibility test of the isolates was performed by Kirby-Bauer disk diffusion test [16]. In brief, each test isolate was swabbed uniformly onto the surface of Mueller-Hinton agar plates. Antibiotic sterile disks including cefotaxime (CTX: 30 μg) and amikacin (AK: 30 μg) were then placed on to the agar surface of the plate. Following incubation, the inhibition zones, in millimeters, were measured in duplicate and scored as sensitive, intermediate, and resistant categories by the critical breakpoints recommended by the Clinical and Laboratory Standards Institute (CLSI) [17].

2.2.2 Preparation of antibiotic stock solution

Standard powder forms of cefotaxime and amikacin were stored at 4°C till usage. The stock solution of each antibiotic was prepared by weighing and consequently dissolving suitable amounts of the antibiotics reaching a concentration of 1000 μg/mL in Mueller-Hinton broth.

2.2.3 Determination of the minimum inhibitory concentration (MIC)

MIC values of antibiotics were determined by the microdilution method following the recommendations of Papich [18]. Stock solutions of antibiotics were prepared and added to the bottom of a 96-well microtiter plate (Nunc Inc., Roskilde, Denmark). 100 mL of this solution was added to the first well of the 96-well plate and serially diluted. 100 mL of an overnight culture of *P. aeruginosa* was added to each well at a final concentration of 5×10^5 CFU/mL (colony-forming units per milliliter). The microtiter plates were incubated at 35°C for 24 h and the MIC determined as the lowest concentration of antibiotics showing no visible bacterial growth.

2.2.4 Test for synergism

The synergistic effect of the antibiotic combinations was detected using a checkerboard dilution assay [19]. The initial concentration of each drug should be fourfold greater than the desired concentration (MIC concentration) and then diluted twofold. In a screw cap test tube, 0.25 mL of broth of each two drugs to be tested was added to 0.5 mL of broth containing a suspension of the organism to be tested to reach the final volume of 1 mL. The inoculum of the bacterial suspension (in 0.5 mL of broth) should be approximately 2×10^5 colony-forming unit (CFU) to produce a final inoculum of 1×10^5 CFU per mL after the addition of an equal volume of the antimicrobial solutions. Each test composed of 36 tubes set horizontally and vertically, 6 rows in one direction contained twofold serial dilutions of antibiotic 1, and 6 rows in the other direction contained twofold serial dilutions of antibiotic 2; two additional rows contained twofold serial dilution of antibiotic 1 or antibiotic 2 alone. The tubes were incubated at 37°C for 24 and 48 h, the tubes were read as those showing turbidity (+) and those showing no turbidity (−). A fractional inhibitory concentration index was used to interpret the results.

2.2.5 Estimation of FIC index

FIC of each agent was calculated by dividing the MIC of the drug in combination by the MIC of the drug alone. The sum of both FICs (ΣFIC = FIC of antibiotic A + FIC of antibiotic B) in each well was used to categorize the combined activity of antimicrobial agents at the given concentrations as synergistic (ΣFIC <= 0.5), partially synergistic (ΣFIC >0.5 and < 1), additive (ΣFIC = 1), indifferent (ΣFIC >1 and < 4), and antagonistic (ΣFIC > = 4) [20].

2.3 Phenotypic characterization of biofilm production

2.3.1 Congo red agar test

Freeman et al. [21] have described a simple qualitative method to detect biofilm production by using a Congo red agar (CRA) medium. CRA medium was prepared with brain heart infusion agar (Oxoid, UK) 37 g/L, sucrose 50g/L, and Congo red indicator (Oxoid, UK) 8 g/L. The first Congo red dye was prepared as a concentrated aqueous solution and autoclaved (121°C for 15 min) separately from the other medium constituents. Then, it was added to the autoclaved brain heart infusion agar with sucrose at 55°C. In this test, the Congo red dye was used as a pH indicator, showing black coloration at pH ranges between 3.0 and 5.2. Plates with the Congo red agar medium were seeded and incubated in an aerobic environment for 24–48 h at 37°C. Isolates were interpreted according to their colony phenotypes. Black colonies with dry constancy and rough surface and edges were suspected as a positive sign of slime formation, while both black colonies with a smooth, round, and shiny surface and red colonies of dry texture and rough edges and surface were suspected as intermediate slime producers. Red colonies with smooth, round, and shiny surfaces were indicators for negative slime formation.

2.3.2 Quantitative detection of biofilm by microtiter plate method

The biofilm assay is performed by using flat-bottom microtiter plates (Techno Plastic Products, Switzerland) as described by O'Toole [22]. *P. aeruginosa* isolates were grown at 37°C in tryptic soy broth (TSB; Oxoid, UK). The bacterial cells were then pelleted at 6000 g for 10 min, and the cell pellets were in 5 mL of fresh medium. The optical densities (ODs) of the bacterial suspensions were measured using a spectrophotometer (Model 6305, Jenway Ltd., Essex, UK) and normalized to an absorbance of 1:00 at 600 nm. The cultures were diluted 1:40 in fresh TSB, and 200 μL of cells were aliquoted into a 96-well polystyrene microtiter plate and inoculated for 24 h at 37°C without agitation. After incubation at 37°C for 24 h, the planktonic cells were aspirated, and the wells were washed three times with sterile phosphate-buffered saline (PBS). The plates were inverted and allowed to dry for an hour at room temperature.

For biofilm quantification, 200 μL of 0.1% aqueous crystal violet solution was added to each well, and the plates were allowed to stand for 15 min. The wells were subsequently washed three times with sterile PBS to wash off the excess crystal violet. Crystal violet bound to the biofilm was extracted with 200 μL of an 80:20 (v/v) mixture of ethyl alcohol and acetone, and the absorbance of the extracted crystal violet was measured at 545 nm on ELISA reader (stat fax 2100, USA). A negative control, crystal violet binding to wells was measured for wells exposed only to the medium with no bacteria. All biofilm assays were performed in triplicate. The interpretation of biofilm production was according to the criteria described by Stepanović et al. [23]. Based on these criteria, optical density cutoff value (ODc) is defined as an average OD of negative control +3 × SD (standard deviation) of the negative control. The ability to produce biofilm of each *P. aeruginosa* isolate was classified according to the following criteria: OD ≤ ODc = not a biofilm producer, ODc < OD ≤ 2x ODc = weak biofilm producer, 2x ODc < OD ≤ 4x ODc = moderate biofilm producer, and 4x ODc < OD = strong biofilm producer.

2.4 Molecular evaluation

2.4.1 DNA extraction

DNA extraction from isolates was performed using the QIAamp DNA Mini Kit (Qiagen, Germany, GmbH) with modifications from the manufacturer's recommendations. Concisely, 10 μL of proteinase K and 200 μL of lysis buffer were added to 200 μL of the sample suspension and incubated at 56°C for 10 min. Then, 200 μL of 100% ethanol was added to the lysate followed by washing and centrifugation according to the manufacturer's recommendations. Nucleic acid was eluted with 100 μL of elution buffer.

2.4.2 PCR amplification of biofilm virulence genes

The obtained DNA was examined for the presence of biofilm in a 25 μL reaction comprising 12.5 μL of EmeraldAmp Max PCR Master Mix (Takara, Japan), 1 μL of each primer of 20 pmol concentration, 4.5 μL of water, and 6 μL of DNA template. The reaction was implemented in an Applied Biosystems 2720 Thermal Cycler for the investigation of the presence of biofilm genes. The properties of all used primers, as well as amplicon length and cycling conditions, were synopsized by Ghadaksaz et al. [24] and listed in **Table 1**.

2.4.3 Analysis of the PCR products

The products of PCR were separated by electrophoresis on 1.5% agarose gel (AppliChem, Germany, GmbH) in 1× TBE buffer at room temperature using gradients of 5 V/cm. For gel analysis, 20 μL of the products were loaded in each gel slot. A GelPilot 100 bp DNA ladder (Qiagen, Germany, GmbH) and GeneRuler 100 bp ladder (Fermentas, Germany) were used to verify the size of fragments. The gel was photographed by a gel documentation system (Alpha Innotech, Biometra), and the data were assessed through computer software.

2.4.4 Quantitative analysis of biofilm gene expression

Biofilm gene expression was analyzed by quantitative real-time PCR (qRT-PCR), and the 16S rRNA housekeeping gene of *Pseudomonas aeruginosa* served as internal control with primer sequence F: GGGGGATCTTCGGACCTCA, R: TCCTTAGAGTGCCCACCCG to normalize the expressional levels between samples. Primers were utilized in a 25 μL reaction containing 12.5 μL of the 2× QuantiTect SYBR Green PCR Master Mix (Qiagen, Germany, GmbH), 0.25 μL of RevertAid Reverse Transcriptase (200 U/μL) (Thermo Fisher), 0.5 μL of each primer of 20 pmol concentration, 8.25 μL of water, and 3 μL of RNA template. The reaction was performed in a Stratagene MX3005P real-time PCR machine with specific conditions mentioned in **Table 2**. To estimate the variation of gene expression on the RNA of the different samples, the Ct of each sample was compared with that of the positive control group according to the "ΔΔCt" method stated by Yuan et al. [25].

2.5 Statistical analysis

Data analysis was performed by SPSS version 22 for windows. A t-test was used to detect statistical differences of the experiments including antibiotic combination treatment versus single antibiotic therapy. Moreover, one-way ANOVA was used for contrasting the influence of these remedies on the fold change of biofilm gene expression. A $P \leq 0.05$ value was suspected as statistically significant.

3. Results

3.1 The recovery rate of isolation and identification

Pseudomonas spp. were isolated from 34 of 100 examined broiler chickens (34%) as shown in **Table 3**. They were further identified by standard microbiological techniques, and an API giving an overall prevalence of 21% was identified as *Pseudomonas aeruginosa*.

Target gene	Primer sequences	Amplified segment (bp)	Primary denaturation	Amplification (35 cycles)			Final extension
				Secondary denaturation	Annealing	Extension	
PslA	TCCCTACCTCAGCAGCAAGC	656	94°C, 5 min	94°C, 30 s	60°C, 40 s	72°C, 45 s	72°C, 10 min
	TGTTGTAGCCGTAGCGTTTCTG						
PelA	CATACCTTCAGCCATCCGTTCTTC	786	94°C, 5 min	94°C, 30 s	60°C, 40 s	72°C, 45 s	72°C, 10 min
	CGCATTCGCCGCACTCAG						
FliC	TGAACGTGGCTACCAAGAACG	180	94°C, 5 min	94°C, 30 s	56.2 °C, 30 s	72°C, 30 s	72°C, 7 min

Table 1.
Primer sequences, target genes, amplicon sizes, and cycling conditions.

Target gene	Reverse transcription	Primary denaturation	Amplification (40 cycles)			Dissociation curve (1 cycle)			Reference
			Secondary denaturation	Annealing	Extension	Secondary denaturation	Annealing	Final denaturation	
16S rRNA	50°C, 30 min	94°C, 15 min	94°C, 15 s	52°C, 30 s	72°C, 30 s	94°C, 1 min	52°C, 1 min	94°C, 1 min	Spilker et al. [26]
pslA				60°C, 30 s			60°C, 1 min		
pelA				60°C, 30 s			60°C, 1 min		Ghadaksaz et al. [24]
fliC				56.2 °C, 30 s			56.2 °C, 1 min		

Table 2.
Target genes and cycling conditions for SYBR green rt-PCR.

Sample	No. of examined samples	*Pseudomonas* spp. isolates		*P. aeruginosa* isolates	
		No.	Frequency	No.	Frequency
Freshly dead	28	11	39%	11	39%
Diseased chicks Young (1–10 days)	33	20	60%	9	27%
Old broilers (11–35 days)	39	3	7.6%	1	2.5%
Total	100	34	34%	21	21%

Table 3.
The incidence of Pseudomonas aeruginosa *isolated from examined samples.*

3.2 Antimicrobial activity

According to the disk diffusion method, 76.2% of isolates were resistant to cefotaxime, 14.3% were intermediate, and 9.5% were sensitive. Regarding amikacin, 57.2% of isolates were resistant, 9.5% were intermediate, and 33.3% were sensitive. Of interest, 57.2% of isolates were resistant to both tested antibiotics.

According to the microdilution assay, the range of cefotaxime MIC was 8–512 μg/mL, while the amikacin MIC range was 1–64 μg/mL as depicted in **Table 4**.

In the checkerboard technique, the interaction between the combination of cefotaxime and amikacin against *Pseudomonas aeruginosa* was predominantly synergistic, although there were few partially synergistic. Thus no growth or turbidity clearly illustrated the extensive activity of aminoglycoside which was enforced by the second drug: cefotaxime resulting in an antibacterial effect. The synergistic activities of the antimicrobial combinations are detailed in **Table 4**. The combination of amikacin and cefotaxime exerted synergetic effect against 17 isolates, and 4 isolates were partially synergistic. FIC index values ranged from 0.18 to 0.75. Statistical analysis of one sample test revealed no significant difference between synergism effects among all isolates indicating strong synergy between both antibiotics where P-value = 0.088. Antagonism was not detected against any isolate in our study.

3.3 Congo red test

About 66.6% of the isolates were positive for biofilm production with varying degrees. Out of 21 *P. aeruginosa* isolates, 19%, 28.6%, and 19% were strong, intermediate, and negative biofilm producers, respectively. The morphology of all types of colonies is illustrated in **Figure 1**.

3.4 Microtiter plate test (MTP)

Biofilm quantification analyses showed that 90.4% of the isolates were biofilm producers, indicating that this technique was more efficient than Congo red agar for the detection of biofilm production. The obtained isolates of this study had the following results for the categories of biofilm production: 9.6% were non-adherent, 33.4% weakly adherent, 42.8% moderately adherent, and 14.2% strongly adherent as shown in **Figure 2**.

A comparison of results obtained by the CRA method versus that of MTP assay is declared in **Table 5**. Out of 21 biofilm *P. aeruginosa* isolates by the CRA method, 19 isolates were positive by the MTP approach but with various levels of

Isolates no.	MIC of CTX	MIC of AK	MIC of CTX in combination	MIC of AK in combination	FIC of CTX	FIC of AK	Σ FIC	Interpretation
1	256	32	32	2	0.125	0.06	0.18	Synergistic
2	8	1	2	0.25	0.25	0.25	0.5	Synergistic
3	32	2	2	1	0.06	0.5	0.56	Partially synergistic
4	128	64	32	16	0.25	0.25	0.5	Synergistic
5	32	64	8	16	0.25	0.25	0.5	Synergistic
6	32	64	8	8	0.25	0.125	0.375	Synergistic
7	64	64	16	4	0.25	0.06	0.31	Synergistic
8	8	4	2	1	0.25	0.25	0.5	Synergistic
9	64	64	16	16	0.25	0.25	0.5	Synergistic
10	128	64	16	16	0.125	0.25	0.375	Synergistic
11	32	64	2	32	0.06	0.5	0.56	Partially synergistic
12	32	64	8	16	0.25	0.25	0.5	Synergistic
13	16	4	2	1	0.125	0.25	0.375	Synergistic
14	128	64	8	8	0.06	0.125	0.18	Synergistic
15	256	64	16	8	0.06	0.125	0.18	Synergistic
16	256	32	64	2	0.25	0.06	0.31	Synergistic
17	32	8	8	4	0.25	0. 5	0.75	Partially synergistic
18	16	4	4	2	0.25	0.5	0.75	Partially synergistic
19	16	4	2	1	0.25	0.25	0.5	Synergistic
20	256	64	64	16	0.25	0.25	0.5	Synergistic
21	512	64	64	8	0.125	0.125	0.25	Synergistic

Table 4.
MIC of CTX and AK alone and in combination and FIC index against P. aeruginosa *by the checkerboard method.*

production (3 strong, 7 moderate, 9 weak), and only 2 isolates were factual negative by both assays.

3.5 Detection of biofilm genes in strong biofilm *P. aeruginosa* isolates by conventional multiplex PCR

All strong biofilm producers *P. aeruginosa* isolates of code numbers (1, 4, 21) were harbored all examined biofilm genes and gave their characteristic bands as shown in **Figure 3**.

3.6 Quantitative assessment effect of each antibiotic alone and in combination on biofilm gene expression

By RT-PCR, comparing the amount of examining biofilm gene products before and after each treatment with a sub-inhibitory concentration (SIC) of each antibiotic alone and combination, results revealed that the amount of examining

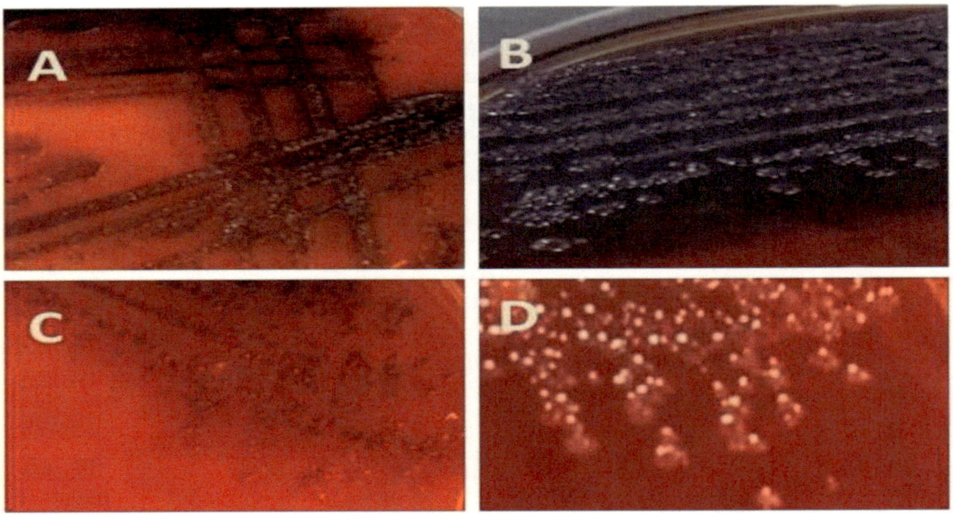

Figure 1.
Investigation of biofilm producer P. aeruginosa *using CRA method: (A) dry black colonies, (B) smooth black colonies, (C) dry red colonies, and (D) smooth red colonies.*

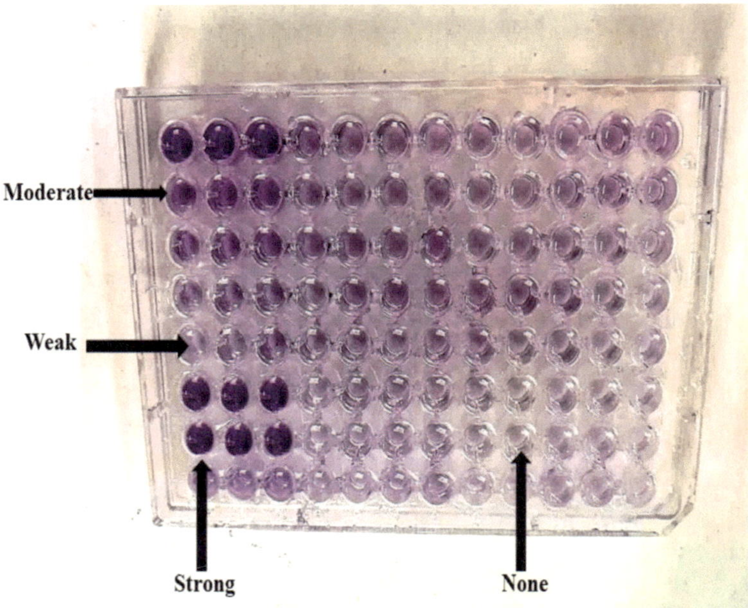

Figure 2.
Microtiter plate method showing none, strong, moderate, and weak biofilm producers differentiated by crystal violet stain in 96-well tissue culture plate.

gene products was relatively increased in untreated samples with drugs than those treated, which leads to high threshold cycle (Ct) value in treated than untreated. Interestingly, we found that drug combination was more effective in significantly reducing the expression of biofilm genes than each antibiotic alone.

Statistical data assessed that fold changes in *pslA*, *pelA*, and *filC* gene expression after treatment with SIC of cefotaxime and amikacin alone were (0.599:0.752:0.597

Sample code no.	CRA	No.	MTP			
			Strong	Moderate	Weak	None
1, 16, 20, 21	Dry black	4	2	2	0	0
4, 7, 9, 10, 14, 15	Smooth black	6	1	3	2	0
3, 5, 6, 17	Dry red	4	0	2	2	0
2, 8, 11, 12, 13, 18, 19	Smooth red	7	0	0	5	2

Table 5.
CRA versus MTP methods for detection of biofilm formation by P. aeruginosa.

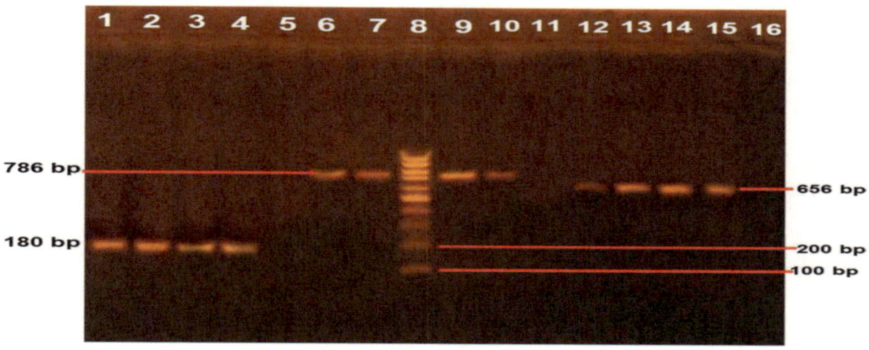

Figure 3.
Agarose gel electrophoresis of biofilm genes: Lanes 1, 6, and 12, positive controls; lanes 5, 11, and 16, negative controls; lane 8, DNA ladder (100 bp); lanes 2–4, positive isolates for filC *gene; lanes 7, 9, and 10, positive isolates for* pelA *gene; and lanes 13–15, positive isolates for* pslA *gene.*

Genes	Isolate no.	Fold change		
		Cefotaxime	Amikacin	Cefotaxime-amikacin combination
PslA	1	0.5212	0.3209	0.0890
	2	0.6830	0.3121	0.1869
	3	0.5946	0.4118	0.1216
PelA	1	0.7371	0.4506	0.2535
	2	0.8526	0.3276	0.2253
	3	0.6690	0.2852	0.1550
FliC	1	0.6071	0.3322	0.2176
	2	0.5471	0.2643	0.1708
	3	0.6373	0.2932	0.0884

Table 6.
Results of RT-PCR showing expression of biofilm genes in P. aeruginosa *isolates before and after treatment with SIC of each antibiotic alone and in combination.*

fold) and (0.348:0.354:0.296 fold), respectively, which were significantly higher ($P \leq 0.05$) than a fold change in same gene expression after combination treatment (0.132:0.211:0.158 fold) as shown in **Table 6** and **Figure 4.**

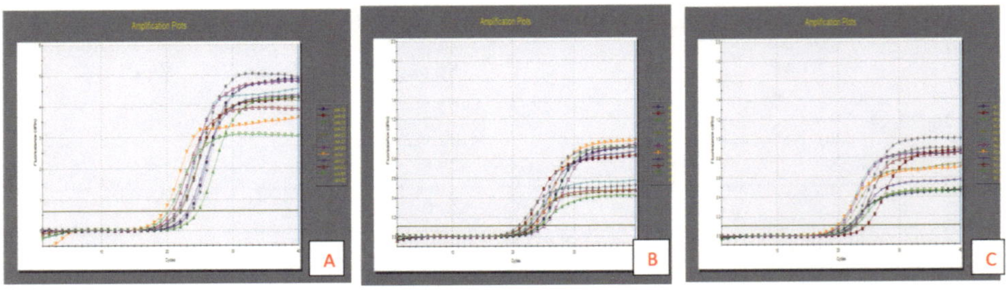

Figure 4.
Expression curves of each biofilm gene after different treatments by SYBR green RT-PCR, (A) PslA gene, (B) PelA gene and (C) FilC gene.

4. Discussion

Pseudomonas aeruginosa is considered to be an opportunistic organism that produces respiratory infection, sinusitis, keratitis/keratoconjunctivitis, and septicemia, and it becomes an infection when it is introduced into tissues of susceptible hosts [27]. The bacterium is widely distributed in the environment, as it can utilize a wide range of materials for its nutrients while only requiring a limited amount of nutrients to survive [28]. Moreover, biofilm production has been considered to be an important determinant of pathogenicity in *P. aeruginosa* infections [29]. The formation of biofilms facilitates chronic bacterial infections and reduces the efficacy of antimicrobial therapy [29–31]. The situation is getting very concerning, the World Health Organization has declared it to be a "critical priority pathogen," on which research and development of novel antibiotics should be focused [32]. For this reason, this work designed to find repositionable candidate's antibiotics against *P. aeruginosa* biofilms, which are disreputable for their intensified drug resistance.

Here we isolated 21 *P. aeruginosa* from 100 broiler chickens suffering from respiratory manifestations (21%). These findings were close to that (20%) reported earlier in India [1]. Many studies showed different prevalence rates of *P. aeruginosa* isolates in broilers worldwide: in Iraq, a low rate of 6% was reported [33], while in Nigeria, a high rate of 75% was reported [34]. These differences in prevalence rates may reflect the considerable disparity in the sampling scheme, sample types, pseudomonas detection protocol, and geographic location.

In the current investigation, all the isolates were tested against cefotaxime and amikacin to determine the antibiotic susceptibility patterns. A high-resistance rate was detected for both antibiotics at which 76.2% were resistant to cefotaxime and 57.2% to amikacin. This might be due to the indiscriminate use of antibiotics in the feed of broiler breeders or other environmental possibilities [35].

The increased observance of multiple resistances (mainly to beta-lactam antibiotics) in pseudomonas isolates is making it increasingly difficult to treat infections caused by this pathogen. Resistance to antimicrobials in pseudomonas strains develops via several mechanisms, including the production of specific enzymes (b-lactamases, enzymes that modify aminoglycosides, for example), changes in cell-membrane permeability, and active efflux systems [36].

Interpretative reading was used to detect the bactericidal activity of each antibiotic against isolates with cefotaxime MICs of 8–512 and amikacin MICs of 1–64. These data are reinforced by findings from other countries, including Kuwait [37], Canada [38], China [39], and the USA [40].

Synergy testing has shown evidence of an interaction of two antibiotics in combination against pseudomonas bacterial isolates where statistical analysis provides important insights into drug synergism where the FIC index calculations exemplified a significant synergism of both drugs achieving an enhanced overall effect which is substantially greater than the sum of their ones. These results were consistent with the previous studies of Saiman [41], Dundar and Otkun [42], and Hawkey et al. [43]. The possible explanation for this synergism is the ability of beta-lactam cefotaxime to penetrate the outer membrane of pseudomonas bacteria which thereby increases the permeability of the bacterium to the aminoglycoside amikacin binding to 30S ribosome inhibiting the protein synthesis, thus leading to a synergistic effect in the in vitro studies [44].

To investigate the effect of a synergistic combination of the repositionable drugs against *P. aeruginosa* biofilms, we detected their effect on the expression of screened biofilm genes.

In this study, biofilm production was examined qualitatively, depending on colony morphology of 21 *P. aeruginosa* isolates inoculated on Congo red agar. Some differences between researches were apparent concerning the interpretation of CRA test results. In that respect, both bright black colonies [45] and black colonies [46] were considered as a positive result. However, Cucarella et al. [47] described the dry crystalline surface (rough colony morphology) as a positive result, disregarding the color (black or pink). Such discrepancy when interpreting the results may be possible since the test itself was not originally designed for investigating *P. aeruginosa* isolates as reported by Freeman et al. [21]. In this investigation, according to Osman et al. [48], isolates that produced black/rough colonies were verified as strong biofilm-forming, while isolates producing red/smooth colonies were described as non-biofilm formers. The smooth black and dry red colonies were respected as indefinite findings.

The qualitative technique revealed that only 66.6% of the isolates were considered biofilm producers, while the biofilm quantitative technique (MTP method) revealed that 90.4% of the isolates were biofilm producers, indicating that the quantitative technique was more efficient than the qualitative technique for the detection of biofilm production. There was also high biofilm production by the evaluated tested isolates of *P. aeruginosa*.

Biofilms are surface-associated communities embedded within an extracellular matrix [49]. The extracellular matrix consists of polysaccharides, proteins, nucleic acids, and lipids and is a distinguishing feature of biofilms, capable of functioning as both a structural scaffold and a protective barrier [45]. Extracellular polysaccharides are a crucial component of the matrix and carry out a range of functions including promoting attachment to surfaces and other cells, building and maintaining biofilm structure, as well as protecting the cells from antimicrobials and host defenses [50, 51].

P. aeruginosa produces at least two extracellular polysaccharides that can be important in biofilm development and is accompanied by gene regulation [52–54].

Conventional PCR was carried out for detection of *pelA* and *pslA* genes which were involved in the formation of polysaccharide components of biofilm among tested isolates and were expressed heavily in all of them (100%). These data matched with previous studies of Wei and Ma [55], Vasiljević et al. [56], and Emami et al. [57].

Moreover, Suriyanarayanan et al. [58] mentioned that the effects of *fliC* phosphorylation on biofilm attachment and dispersal led to two conclusions. Both initial attachment and detachment during the dispersal stage were delayed by the loss of *fliC* phosphorylation in static and dynamic flow biofilms. As each of these processes still proceeded in the lack of phosphorylation, it suggested that *fliC* phosphorylation regulates the timing and rate of these processes without affecting biofilm architecture. These investigations were parallel with our results where *fliC* detected in all tested isolates.

Regarding the qRT-PCR results, the suppressing effects in fold change of previously mentioned biofilm gene expression were detected for drug combination in comparison with each antibiotic alone. Exposure to each antibiotic caused a decreased level of biofilm expression ranging between 0.1- and 0.7-fold changes, while the repression was strong and most significant with amikacin-cefotaxime combination treatment with fold change reaching 0.08, i.e., the consequence of treatment on the average expression profile among all biofilm involving genes constituting the bacterial communities studied. As described in this paper and by others [59–61], sub-MICs of combinations have potent effects on attenuating biofilm formation which are totally different from each antibiotic alone.

5. Conclusion

The treatment of biofilm-related *P. aeruginosa* infections in the poultry industry has become an important part of antimicrobial chemotherapy because biofilms are not affected by therapeutic concentrations of antibiotics permitting attachment of other pathogens. Our study proved that using a combination of antimicrobial agents including cefotaxime and amikacin represents a profound synergism, significant antibiofilm, and a suitable candidate in combatting this fierce infection.

Conflict of interest

The authors manifested that they have no conflicts of interest.

Abbreviations

FIC	fractional inhibitory concentration
MIC	minimum inhibitory concentration
CRA	Congo red agar
MTP	microtiter plate
RT-PCR	reverse transcriptase-polymerase chain reaction

Author details

Azza S. El-Demerdash[1*] and Neveen R. Bakry[2]

1 Agriculture Research Center (ARC), Animal Health Research Institute (AHRI), Zagazig, Egypt

2 Agriculture Research Center (ARC), Animal Health Research Institute (AHRI), Reference Laboratory for Veterinary Quality Control on Poultry Production (RLQCP), Dokki, Egypt

*Address all correspondence to: dr.azzasalah@yahoo.com

References

[1] Shukla S, Mishra P. *Pseudomonas aeruginosa* infections in broiler chicks in Jabalpur. International Journal of Environmental Research. 2015;**6**:37-39. Available from: https://www.researchgate.net/publication/304704794_Pseudomonas_aeruginosa_Infection_in_Broiler_Chicks_in_Jabalpur

[2] Kebede F. Pseudomonas infection in chickens. Journal of Veterinary Medicine and Animal Health. 2010;**2**:55-58. Available from: https://academicjournals.org/journal/JVMAH/article-full-text-pdf/663013539683

[3] Wilkinson GF, Pritchard K. In vitro screening for drug repositioning. Journal of Biomolecular Screening. 2015;**20**:167-179. DOI: 10.1177/1087057114563024

[4] Dandachi I, Sokhn ES, Dahdouh EA, Azar E, El-Bazzal B, Rolain JM, et al. Prevalence and characterization of multi-drug-resistant gram-negative bacilli isolated from lebanese poultry: A nationwide study. Frontiers in Microbiology. 2018;**9**:1-11. DOI: 10.3389/fmicb.2018.00550

[5] Shrivastava SR, Shrivastava PS, Ramasamy J, et al. World health organization releases global priority list of antibiotic-resistant bacteria to guide research, discovery, and development of new antibiotics. Journal of Medical Society. 2018;**32**:76. Available from: http://www.jmedsoc.org/temp/JMedSoc32176-3543534_095035.pdf

[6] Siles SA, Srinivasan A, Pierce CG, Lopez-Ribot JL, Ramasubramanian AK. High-throughput screening of a collection of known pharmacologically active small compounds for identification of *Candida albicans* biofilm inhibitors. Antimicrobial Agents and Chemotherapy. 2013;**57**:3681-3687. DOI: 10.1128/AAC.00680-13

[7] Rangel-Vega A, Bernstein LR, Mandujano Tinoco E-A, Garcia-Contreras S-J, Garcia-Contreras R. Drug repurposing as an alternative for the treatment of recalcitrant bacterial infections. Frontiers in Microbiology. 2015;**6**:282. DOI: 10.3389/fmicb.2015.00282

[8] Torres NS, Abercrombie JJ, Srinivasan A, Lopez-Ribot JL, Ramasubramanian AK, Leung KP. Screening a commercial library of pharmacologically active small molecules against *Staphylococcus aureus* biofilms. Antimicrobial Agents and Chemotherapy. 2016;**60**:5663-5672. DOI: 10.1128/AAC.00377-16

[9] Yssel AEJ, Vanderleyden J, Steenackers HP. Repurposing of nucleoside-and nucleobase-derivative drugs as antibiotics and biofilm inhibitors. The Journal of Antimicrobial Chemotherapy. 2017;**72**:2156-2170. DOI: 10.1093/jac/dkx151

[10] Delattin N, De Brucker K, Vandamme K, Meert E, Marchand A, Chaltin P, et al. Repurposing as a means to increase the activity of amphotericin B and caspofungin against *Candida albicans* biofilms. The Journal of Antimicrobial Chemotherapy. 2013;**69**:1035-1044. DOI: 10.1093/jac/dkt449

[11] den Driessche F, Brackman G, Swimberghe R, Rigole P, Coenye T. Screening a repurposing library for potentiators of antibiotics against Staphylococcus aureus biofilms. International Journal of Antimicrobial Agents. 2017;**49**:315-320. DOI: 10.1016/j.ijantimicag.2016.11.023

[12] Nazli E, Zer Y, Eksi F. In vitro efficacy of various antibiotic combinations against *Pseudomonas aeruginosa* isolates. The Journal of International Medical Research. 2015;**43**:217-225. DOI: 10.1177/0300060514553490

[13] Batoni G, Maisetta G, Esin S. Antimicrobial peptides and their interaction with biofilms of medically relevant bacteria. Biochimica et Biophysica Acta - Biomembranes. 2016;**1858**:1044-1060. DOI: 10.1016/j.bbamem.2015.10.013

[14] Grassi L, Maisetta G, Esin S, Batoni G. Combination strategies to enhance the efficacy of antimicrobial peptides against bacterial biofilms. Frontiers in Microbiology. 2017;**8**:2409. Available from: https://www.ncbi.nlm.nih.gov/pmc/articles/PMC5770624/

[15] Health protection agency. Identification of pseudomonas species and other non-glucose fermenters. Standards for microbiology investigations (UK SMI). 2015;**3**:3-41. Available from: https://assets.publishing.service.gov.uk/government/uploads/system/uploads/attachment_data/file/422699/ID_17i3.pdf

[16] Jorgensen JH, Turnidge JD. Susceptibility test methods: Dilution and disk diffusion methods. In: Manual of Clinical Microbiology, 11th ed. American Society of Microbiology; 2015. pp. 1253-1273. Available from: https://www.asmscience.org/content/book/10.1128/9781555817381.mcm11.ch71?crawler=true

[17] CLSI (Clinical and Laboratory Standards Institute). M100-S25 Performance Standards for Antimicrobial. 2015;**35**(3):1-240. Available from: http://file.qums.ac.ir/repository/mmrc/CLSI2015.pdf

[18] Papich MG. Antimicrobials, susceptibility testing, and minimum inhibitory concentrations (MIC) in veterinary infection treatment. The Veterinary Clinics of North America. Small Animal Practice. 2013;**43**:1079-1089. DOI: 10.1016/j.cvsm.2013.04.005

[19] Kim S-H, Park C, Chun H-S, Lee D-G, Choi J-K, Lee H-J, et al. Pilot screening to determine antimicrobial synergies in a multidrug-resistant bacterial strain library. Microbial Drug Resistance. 2016;**22**:372-378. Available from: https://www.ncbi.nlm.nih.gov/pmc/articles/PMC4939386/

[20] Orhan G, Bayram A, Zer Y, Balci I. Synergy tests by E test and checkerboard methods of antimicrobial combinations against Brucella melitensis. Journal of Clinical Microbiology. 2005;**43**:140-143. DOI: 10.1128/JCM.43.1.140-143.2005

[21] Freeman DJ, Falkiner FR, Keane CT. New method for detecting slime production by coagulase negative staphylococci. Journal of Clinical Pathology. 1989;**42**:872-874. DOI: 10.1136/jcp.42.8.872

[22] O'Toole GA. Microtiter dish biofilm formation assay. Journal of Visualized Experiments. 2011;**47**:1-2. DOI: 10.3791/2437

[23] Stepanović S, Vuković D, Hola V, Bonaventura GDI, Djukić S, Ćirković I, et al. Quantification of biofilm in microtiter plates: Overview of testing conditions and practical recommendations for assessment of biofilm production by staphylococci. APMIS. 2007;**115**:891-899. DOI: 10.1111/j.1600-0463.2007.apm_630.x

[24] Ghadaksaz A, Fooladi AAI, Hosseini HM, Amin M. The prevalence of some pseudomonas virulence genes related to biofilm formation and alginate production among clinical isolates. Journal of Applied Biomedicine. 2015;**13**:61-68. DOI: 10.1016/j.jab.2014.05.002

[25] Yuan JS, Reed A, Chen F, Stewart CN. Statistical analysis of real-time PCR data. BMC Bioinformatics. 2006;7:85. DOI: 10.1186/1471-2105-7-85

[26] Spilker T, Coenye T, Vandamme P, LiPuma JJ. PCR-based assay for differentiation of *Pseudomonas aeruginosa* from other *Pseudomonas*

species recovered from cystic fibrosis patients. Journal of Clinical Microbiology. 2004;**42**:2074-2079. DOI: 10.1128/jcm.42.5.2074-2079.2004

[27] Goldberg JB. Why is *Pseudomonas aeruginosa* a pathogen? F1000 Biology Reports. 2010;**2**:29. DOI: 10.3410/B2-29

[28] Abdelraouf K, Kabbara S, Ledesma KR, Poole K, Tam VH. Effect of multidrug resistance-conferring mutations on the fitness and virulence of Pseudomonas aeruginosa. The Journal of Antimicrobial Chemotherapy. 2011;**66**:1311-1317. DOI: 10.1093/jac/dkr105

[29] Wareham DW, Curtis MA. A genotypic and phenotypic comparison of type III secretion profiles of *Pseudomonas aeruginosa* cystic fibrosis and bacteremia isolates. International Journal of Medical Microbiology. 2007;**297**:227-234. DOI: 10.1016/j.ijmm.2007.02.004

[30] Parsek MR, Singh PK. Bacterial biofilms: An emerging link to disease pathogenesis. Annual Review of Microbiology. 2003;**57**:677-701. DOI: 10.1146/annurev.micro.57.030502.090720

[31] Finnan S, Morrissey JP, O'gara F, Boyd EF. Genome diversity of *Pseudomonas aeruginosa* isolates from cystic fibrosis patients and the hospital environment. Journal of Clinical Microbiology. 2004;**42**:5783-5792. DOI: 10.1128/JCM.42.12

[32] Tacconelli E, Carrara E, Savoldi A, Harbarth S, Mendelson M, Monnet DL, et al. Discovery, research, and development of new antibiotics: The WHO priority list of antibiotic-resistant bacteria and tuberculosis. The Lancet Infectious Diseases. 2018;**18**:318-327. DOI: 10.1016/S1473-3099(17)30753-3

[33] Noori TE, Alwan MJ. Isolation and identification of zoonotic bacteria from poultry meat. International Journal of Advanced Research in Biological Sciencesi. 2016;**3**:57-66. Available from: https://pdfs.semanticscholar.org/488e/4253f660528b486642056c4d74aa3b6cc742.pdf

[34] Abiala M, Olayiwola J, Babatunde O, Aiyelaagbe O, Akinyemi S. Evaluation of therapeutic potentials of plant extracts against poultry bacteria threatening public health. BMC Complementary and Alternative Medicine. 2016;**16**:417. DOI: 10.1186/s12906-016-1399-z

[35] Nazer AHK, Dadras H, Eskandari S. Aerobic bacteria isolated from eggs and day-old chicks and their antibacterial resistance in shiraz, Iran. Iranian Journal of Veterinary Research. 2006;**7**:20-30. Available from: https://www.sid.ir/en/Journal/ViewPaper.aspx?ID=43085

[36] Nathwani D, Raman G, Sulham K, Gavaghan M, Menon V. Clinical and economic consequences of hospital-acquired resistant and multidrug-resistant *Pseudomonas aeruginosa* infections: A systematic review and meta-analysis. Antimicrobial Resistance and Infection Control. 2014;**3**:32. DOI: 10.1186/2047-2994

[37] Alfouzan W, Dhar R, Nicolau D. In vitro activity of newer and conventional antimicrobial agents, including Fosfomycin and Colistin, against selected gram-negative bacilli in Kuwait. Pathogens. 2018;**7**:75. DOI: 10.3390/pathogens7030075

[38] Domalaon R, Sanchak Y, Koskei LC, Lyu Y, Zhanel GG, Arthur G, et al. Short proline-rich lipopeptide potentiates minocycline and rifampin against multidrug-and extensively drug-resistant *Pseudomonas aeruginosa*. Antimicrobial Agents and Chemotherapy. 2018;**62**:e02374-e02317

[39] Kuti JL, Wang Q, Chen H, Li H, Wang H, Nicolau DP. Defining the potency of amikacin against *Escherichia coli, Klebsiella pneumoniae, Pseudomonas*

aeruginosa, and *Acinetobacter baumannii* derived from Chinese hospitals using CLSI and inhalation-based breakpoints. Infection and Drug Resistance. 2018;**11**:783. DOI: 10.2147/IDR.S161636

[40] Sutherland CA, Verastegui JE, Nicolau DP. In vitro potency of amikacin and comparators against *E. coli, K. pneumoniae* and *P. aeruginosa* respiratory and blood isolates. Annals of Clinical Microbiology and Antimicrobials. 2016;**15**:39. Available from: https://www.ncbi.nlm.nih.gov/pmc/articles/PMC4912699/

[41] Saiman L. Clinical utility of synergy testing for multidrug-resistant *Pseudomonas aeruginosa* isolated from patients with cystic fibrosis. Paediatric Respiratory Reviews. 2007;**8**:249-255. Available from: https://www.ncbi.nlm.nih.gov/pubmed/17868923

[42] Dundar D, Otkun M. In-vitro efficacy of synergistic antibiotic combinations in multidrug resistant *Pseudomonas aeruginosa* strains. Yonsei Medical Journal. 2010;**51**:111-116. DOI: 10.3349/ymj.2010.51.1.111

[43] Hawkey PM, Warren RE, Livermore DM, McNulty CAM, Enoch DA, Otter JA, et al. Treatment of infections caused by multidrug-resistant gram-negative bacteria: Report of the British Society for Antimicrobial Chemotherapy/healthcare infection society/british infection association joint working party. The Journal of Antimicrobial Chemotherapy. 2018;**73**:iii2-iii78

[44] Spoorthi NJ, Vishwanatha T, Reena V, Divyashree BC, Aishwarya S, Siddhalingeshwara KG, et al. Antibiotic synergy test: Checkerboard method on multidrug resistant *Pseudomonas aeruginosa*. International Research Journal of Pharmacy. 2011;**2**:196-198. Available from: https://irjponline.com/admin/php/uploads/754_pdf.pdf

[45] Häussler S, Parsek MR. Biofilms 2009: new perspectives at the heart of surface-associated microbial communities 2010. DOI: 10.1128/JB.00332-10

[46] Jain A, Agarwal A. Biofilm production, a marker of pathogenic potential of colonizing and commensal staphylococci. Journal of Microbiological Methods. 2009;**76**: 88-92. Available from: https://www.ncbi.nlm.nih.gov/pubmed/18851996

[47] Cucarella C, Tormo MA, Ubeda C, Trotonda MP, Monzón M, Peris C, et al. Role of biofilm-associated protein bap in the pathogenesis of bovine *Staphylococcus aureus*. Infection and Immunity. 2004; **72**:2177-2185. Available from: https://www.ncbi.nlm.nih.gov/pubmed/15039341

[48] Osman K, Alvarez-Ordóñez A, Ruiz L, Badr J, ElHofy F, Al-Maary KS, et al. Antimicrobial resistance and virulence characterization of *Staphylococcus aureus* and coagulase-negative staphylococci from imported beef meat. Annals of Clinical Microbiology and Antimicrobials. 2017;**16**:1-10. DOI: 10.1186/s12941-017-0210-4

[49] Rasamiravaka T, Vandeputte OM, Pottier L, Huet J, Rabemanantsoa C, Kiendrebeogo M, et al. *Pseudomonas aeruginosa* biofilm formation and persistence, along with the production of quorum sensing-dependent virulence factors, are disrupted by a triterpenoid coumarate ester isolated from Dalbergia trichocarpa, a tropical legume. PLoS One. 2015;**10**:e0132791

[50] Smith-Palmer A, Stewart J, Fyfe L. The potential application of plant essential oils as natural food preservatives in soft cheese. Food Microbiology. 2001;**18**:463-470. Available from: https://www.sciencedirect.com/science/article/pii/S0740002001904150

[51] Stoodley P, Sauer K, Davies DG, Costerton JW. Biofilms as complex differentiated communities. Annual Review of Microbiology. 2002;**56**: 187-209. Available from: https://www.ncbi.nlm.nih.gov/pubmed/12142477

[52] Costerton JW, Stewart PS, Greenberg EP. Bacterial biofilms: A common cause of persistent infections. Science. 1999;**284**:1318-1322. Available from: https://www.ncbi.nlm.nih.gov/pubmed/10334980

[53] Drenkard E. Antimicrobial resistance of Pseudomonas aeruginosa biofilms. Microbes and Infection. 2003;**5**:1213-1219. Available from: https://www.ncbi.nlm.nih.gov/pubmed/14623017

[54] Wei Q, Ma LZ. Biofilm matrix and its regulation in *Pseudomonas aeruginosa*. International Journal of Molecular Sciences. 2013;**14**:20983-21005. Available from: https://www.ncbi.nlm.nih.gov/pubmed/24145749

[55] Wei B, Cha S-Y, Kang M, Park I-J, Moon O-K, Park C-K, et al. Development and application of a multiplex PCR assay for rapid detection of 4 major bacterial pathogens in ducks. Poultry Science. 2013;**92**:1164-1170. Available from: https://academic.oup.com/ps/article/92/5/1164/1573914

[56] Vasiljević Z, Jovčić B, Ćirković I, DJukić S. An examination of potential differences in biofilm production among different genotypes of *Pseudomonas aeruginosa*. Archives of Biological Sciences. 2014;**66**:117-121. Available from: http://www.doiserbia.nb.rs/Article.aspx?ID=0354-46641401117V#.XhmKw9QzbIU

[57] Emami S, Nikokar I, Ghasemi Y, Ebrahimpour M, Ebrahim-Saraie HS, Araghian A, et al. Antibiotic resistance pattern and distribution of pslA gene among biofilm producing *Pseudomonas aeruginosa* isolated from waste water of a burn center. Jundishapur Journal of Microbiology. 2015;**8**(11):e23669. DOI: 10.5812/jjm.23669

[58] Suriyanarayanan T, Periasamy S, Lin MH, Ishihama Y, Swarup S. Flagellin FliC phosphorylation affects type 2 protease secretion and biofilm dispersal in *Pseudomonas aeruginosa* PAO1. PLoS One. 2016;**11**(10):1-19. DOI: 10.1371/journal.pone.0164155

[59] Gümücs D, Kalayci-Yüksek F, Yörük E, Uz G, Çelik E, Arslan C, et al. Alterations of growth rate and gene expression levels of UPEC by antibiotics at sub-MIC. Folia Microbiologia (Praha). 2018;**63**:451-457. Available from: https://www.ncbi.nlm.nih.gov/pubmed/29327292

[60] Raymond B. Five rules for resistance management in the antibiotic apocalypse, a road map for integrated microbial management. Evolutionary Applications. 2019;**12**(6):1079-1091. DOI: 10.1111/eva.12808

[61] Shahat HS, Mohamed H, Al-Azeem A, Mohammed W, Nasef SA. Molecular detection of some virulence genes in *Pseudomonas aeruginosa* isolated from chicken embryos and broilers with regard to disinfectant resistance. SVU-International Journal of Veterinary Sciences. 2019;**2**:52-70. Available from: https://svu.journals.ekb.eg/article_40727_bc49d8a75991c214123ab7c8573d73a4.pdf

Chapter 8

Gene Expression Profile of HDF in SMG Partially Overlaps with That in the NASA Twins Study

Jade Q. Clement

Abstract

Microgravity research is an important field in biomedical sciences not only due to our interest in exploring and living in space, but also because of the insights it gives on earthbound health conditions. Using a human dermal fibroblast (HDF) cell line cultured in simulated microgravity (SMG) in combination with high throughput cDNA microarrays and quantitative Northern analysis, 271 differentially regulated genes were identified and 72% of these genes were also reported in the high throughput gene expression data of the recent National Aeronautics and Space Administration (NASA) Twins Study. The identification of the large number of overlapping microgravity sensitive genes between the skin fibroblast in microgravity and astronaut's peripheral blood mononuclear cells (PBMCs) indicated that microgravity alone, without space radiation, was able to elicit an adaptive response involving a set of about 200 genes. Further analysis of the overlapping genes with the same direction of regulation (86 genes) and opposite direction of regulation (108 genes) revealed important pathways and cellular processes in the microgravity adaptation responses.

Keywords: gene expression profiling, DNA microarray, northern blotting, rotating wall vessel (RWV), rotary cell culture system (RCCS), fibroblast, microgravity

1. Introduction

Humans have been traveling to space since 1961. During the close to 60-year period, hundreds of astronauts and cosmonauts have experienced microgravity as well as radiation in space. Exposure to microgravity environment has been shown to have potentially negative effects on human health. Some of the factors include cardiovascular deconditioning [1, 2], decline in immune response [3, 4], bone deterioration [5–7], and muscular atrophy [8–10].

Human immune response dysregulation has been shown during and after space flight [11]. Although acute onset life-threatening incidents directly caused by microgravity or spaceflight have not been reported, many conditions of health concern and symptoms related to immune dysfunction during orbital spaceflight have been described [12, 13]. Among the 70 tabulated clinical symptoms and medical conditions pertinent to immune dysfunction on board the International Space Station (ISS), skin rash and hypersensitivity account for 23 events; skin

IntechOpen

infections, 6 events; cold sores (caused by herpes simplex virus 1 infection), 6 events. About 50% of the tabulated incidents are of or closely related to skin symptoms or abnormal skin conditions during the long duration space flights [12]. Skin is the essential outer cover that protects the internal tissues and organs from potential physical, chemical, and biological assaults of the environment. In addition to immune cells and keratinocytes, dermal fibroblasts also play an important immunomodulation role such as in antimicrobial defense [14]. Epidermal keratinocytes can sense the presence of pathogen invasion and other environmental stimuli such as the presence of UV light and foreign chemicals and produce cytokines, chemokines, and growth factors in response. Communication between keratinocytes and dermal fibroblasts through cytokines is fundamental in skin immunity. A recent report shows that in a 3D skin model-based study, just keratinocytes and fibroblasts alone embedded in a collagen matrix are able to activate CD4+ T cells in response to microbial invasion [15]. The dermal fibroblasts play an essential role in antimicrobial response by integrating signals among cells in the skin.

To date, much microgravity research work has been done in ground-based research using microgravity analogs. Due to cost and limits to the technology much less has been done directly in the space environment. The recent NASA Twins Study is a tremendously important study because it is the first study that uses an integrated approach to study human adaptation to a space environment by documenting the molecular, physiological and cognitive effects during long term spaceflight [16]. The study also highlights the need for further study on important aspects such as vascular changes and immunological stress associated with the weightlessness of space flight [16].

Sudden gravity change has altered gene expressions from many cell types [17–19]. High-throughput gene expression analysis have great potential for application to research involving changes in environmental conditions [20]. Various high throughput studies such as cDNA microarrays and transcriptome RNA sequencing have been increasingly used to assess the mRNA levels in microgravity research [16, 17]. This is an effective approach because the control of mRNA abundance of genes is efficiently adapted by cells through controlling transcription (especially transcription initiation), nuclear pre-mRNA processing, mRNA transport, mRNA stability, etc. The cellular abundance of mRNAs is critical to gene function and protein production which are intriguingly fine-tuned by non-coding regulatory RNAs such as miRNAs. There have been many gene expression studies done on various cell lines grown both in space and using ground-based microgravity analogs. Many of these studies have yielded valuable data, but correlation of gene expression data between studies has been relatively low [17, 18].

To further understand the cellular and molecular mechanisms by which space flight alters skin immune defense activities such as those analyzed from the ISS [12], the effects of microgravity on various human skin cell lines need to be studied to identify the genes whose functions are altered by microgravity. In our previous study, we found expression changes in certain genes (such as HLA-G and IL-1β among many others) in response to simulated microgravity [21]. The current report is on the gene expression profile of HDF in response to SMG. Interestingly, a substantial overlap in gene expression profiles between the HDF under SMG and that from the human blood cells of the NASA Twins Study, especially the peripheral blood mononuclear cells (PBMCs) from inflight in the ISS. The comparative analysis yielded 194 differently expressed genes in both studies, of which 86 genes were regulated in the same direction (trend) while 108 genes were regulated oppositely. The significance of these findings was discussed.

2. Materials and methods

2.1 Simulated microgravity and cell culture

The HDF cell line, AG 1522, was generously provided by Dr. Honglu Wu of NASA. The HDF cell line displayed regular monolayer spindle shaped growth in conventional 2-D cell culture flasks. When HDF cells were subjected to SMG treatment, in a 3-D culture environment, they formed spherical aggregates. Ground based simulated microgravity was achieved using the 50 ml high aspect ratio vessels (HARVs) or rotating wall vessels (RWVs) of a rotary cell culture system (RCCS-4D) bioreactors from Synthecon, Inc. Cell viability and cell concentration were determined by Vi-Cell 1.01 cell counter of Beckman Coulter. For the three parallel experiments of 5 day modeled microgravity exposures, a density of 2.0×10^6 cells/ml with viability of 95.5% of the HDF AG1522 cells were cultured in RWVs at 20 rpm rotary setting to achieve the constant free-fall experience for cell aggregates. At the end of the five-day microgravity exposure period, the content of the bioreactor vessels was poured out into a 50 ml sterile centrifuge tube to collect cell pellet and 5 ml of the cell suspension from the bioreactor vessels were transferred to T75 flasks for morphological observation. Non-exposed stationary normal gravity control AG1522 cells were cultured in tissue culture flasks with vented caps (TPP Techno Plastic) in the same incubator at 37°C, 5% CO_2.

2.2 Total RNA isolation and DNA microarray hybridization

HDF cells cultured in three SMG bioreactor vessels and control flasks were removed at the end of the five-day SMG exposure period, washed with phosphate buffered saline three times and lysed in Guanidinium Isothiocyanate Buffer. The cell lysates were stored at −80°C prior to ultracentrifugation for total RNA isolation [22, 23]. Total cellular RNA was labeled using the Agilent Low RNA Input Fluorescent Linear Amplification Kit following manufacturer's protocols [24]. The fluorescently labeled cRNA probes were further purified and hybridized to Agilent 22 K Human Microarray V2 according to the specified procedures within the kit.

2.3 Microarray scanning, feature extraction and functional grouping

The microarrays were scanned using a ScanArray microarray scanner (Perkin-Elmer). The images generated from the scanning were imported into GenePix 6.0 (Molecular Devices, Sunnyvale, CA) for alignment and initial quantitation. The Gene Pix Results (GPR) files were then uploaded to CARMAweb [25] for normalization and statistical analysis. Background was subtracted and then each array was normalized using loess normalization within the array. A paired moderated T-Test was applied with Benjamini-Hochberg correction to control the false discovery rate [26]. Cut-offs were set at a P-value ≤ 0.01 and fold change of ≥ 1.5.

2.4 Northern blotting and quantitative gene expression analysis

Some of the significantly regulated microgravity sensitive genes identified from the DNA microarray analysis were further verified using Northern blotting. Briefly, 10 ug total RNA was loaded per lane on a 1% formaldehyde agarose gel for electrophoresis separation of RNA species. RNA Ladders from Fermentas Life Sciences were used as RNA size markers. The gel-separated RNAs were capillary transferred onto a nylon membrane which was subjected to a hybridization procedure using chemiluminescent (Pierce Biotechnology) labeled cDNA probe fragments. The blot was sequentially hybridized and striped and hybridized again with 12 different probe fragments. The cDNA probe fragments were generated from reverse transcription polymerase chain reaction (RT-PCR) using total cellular RNA as the templates. RT-PCRs were carried out using the Reverse Transcription System of Promega and the BioLine Red Polymerase PCR kit. Northern blot quantitation and association with the cDNA array results were done as described previously [21, 23].

2.5 Comparative analysis with the data from the NASA twins study

The NASA Twin gene expression data was identified from Suppemental Table 2 in the NASA Twins Study [16]. Prior to further bioinformatics analysis, the high throughput gene expression data from transcriptome RNA sequencing analysis for the inflight, first half and in-flight, second half was extracted from the NASA Twin study [16]. This data was then compared to the gene expression data from the current study with HDF in SMG. Mainly the transcriptome RNA sequencing data of the PBMC RNA from NASA Twins Study was compared with the HDF SMG data here, since it offered the most abundant overlapping genes. The PBMC data was extracted from the Excel spreadsheets and combined with the HDF data for further comparative analysis using similar previously published method [17].

2.6 Pathway and gene ontology analysis

To determine the Kyoto Encyclopedia of Genes and Genomes (KEGG) Pathways [27, 28] and Gene Ontology, the genes that were determined to be differentially regulated were uploaded to the Database for Annotation, Visualization and Integrated Discovery (DAVID) [29]. The pathway and gene ontology information were used to build the tables. The process was done for both the HDF SMG data alone, and for its comparative analysis with the data from the NASA Twins Study. Prior to further bioinformatics analysis, the gene expression data for the inflight, first half and in-flight, second half was extracted from Supplemental Table 2 of the NASA Twins study [16]. This data was then compared to the gene expression data from the current study with HDF in SMG.

2.7 Constructing heat map and Venn diagram

The heat map was generated using Genesis 1.8.1 [30]. The Venn diagram was generated using the Excel plugin Array File Maker 4.0 [31].

3. Results

3.1 Gene expression profiling and the identification of microgravity sensitive genes from SMG treated HDF

HDF cells from each of the three SMG bioreactors and normal gravity controls were removed after 5-day SMG exposure for RNA extraction and microarray experiments. After normalization and statistical analysis (student *t* test), the gene expression data were used to identify the initial set of significantly differentially regulated genes at the statistically significance level of $P \leq 0.01$ and cut off point of ≥ 1.5 fold up or down regulation. The volcano plot (**Figure 1**) shows the overall profile of the gene expression data from the three sets of parallel SMG experiments. Each dot on the volcano plot represents a gene selected in the initial set of differentially

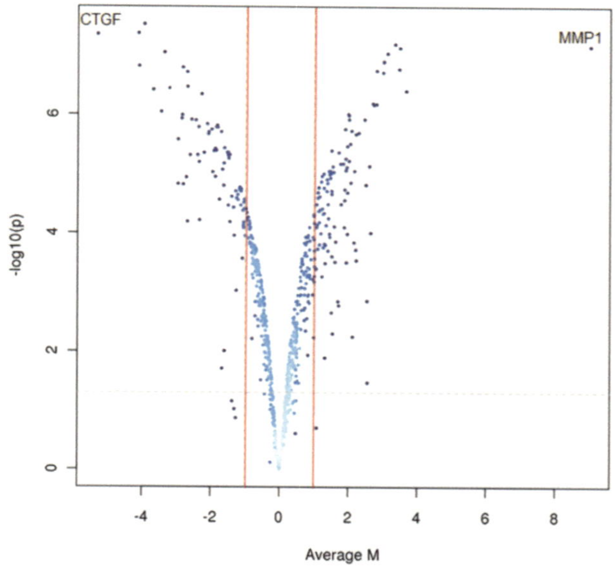

Figure 1.
Volcano plot scattering the average M values (x axis) against the corrected p values (y axis) created using CARMAweb tool. Dots to the left of the red line and above the dashed line have a p-value of 0.01 or less and are down-regulated by at least 1.5 fold. The dots to the right of the red line and above the dashed lines have a p-value of 0.01 or less and are up-regulated by at least 1.5 fold.

expressed ones if the dot is to the left (down) or right (up) of the red lines. The genes found outside of the vertical parallel red lines correspond to the genes differentially expressed by 1.5 fold or greater in the SMG. A total of 271 genes were identified from these three SMG experiments on HDF to be the initial set of significant genes that were used for further analysis as follows. The two most notable differentially regulated genes by the SMG evident on the plot were the matrix metallopeptidase 1 (MMP1) and connective tissue growth factor (CTGF) genes (**Figure 1**).

The significantly (≥ 1.5 fold change, $P \leq 0.01$) differentially expressed genes under simulated microgravity were further analyzed using Heatmap and Venn diagram to visually display the directions and centralization of gene expression. A high level of consistency of both the Heatmap and Venn diagram results were found among the microarray data from the three SMG bioreactors (**Figure 2**). The Heatmap indicated that the expression levels of the three replicates were similar, with very minor variations in magnitude of expression. The Venn diagram further showed that there was a complete match among the three replicate microarrays from the three SMG bioreactor RNA samples. Thus, there was a very high level of consistence among data from the three SMG bioreactors. Among the 271 microgravity sensitive genes that differentially regulated by 1.5 fold or greater with a P value of ≤ 0.01, 129 were down-regulated and 142 were upregulated (**Figure 2B**).

3.2 KEGG pathways of the microgravity sensitive genes from HDF

The identified 271 SMG sensitive genes were subjected to further bioinformatics analysis using the DAVID v6.8, which uses a modified Fisher Exact P-value for gene enrichment analysis and statistically determines the over-representation of functional gene categories in a gene list. P values equal to or smaller than 0.05 are considered significantly enriched [29]. Through the DAVID analysis, 16 statistically significant ($p \leq 0.05$) KEGG Pathways from the SMG gene list were identified and genes in all the 23 KEGG pathways had >2 fold enrichment (**Table 1**).

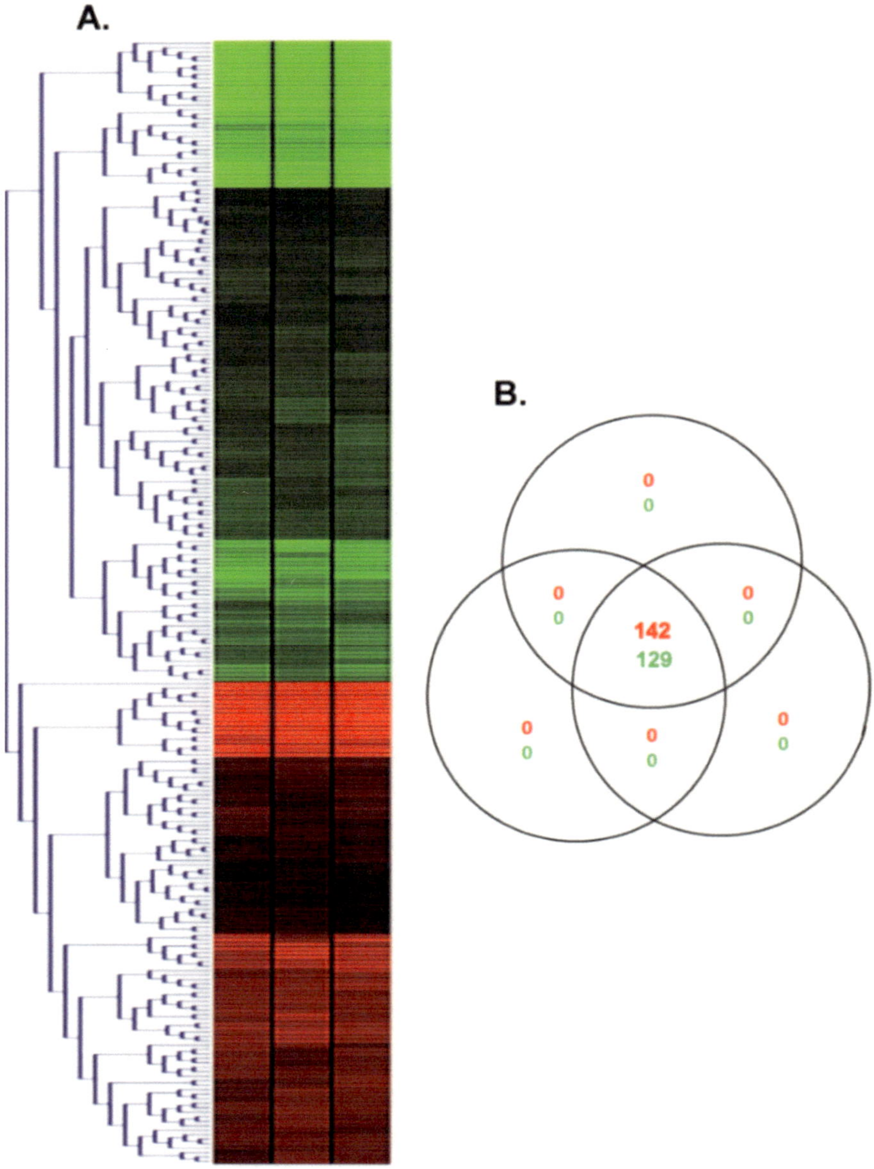

Figure 2.
Heatmap and Venn diagram comparing the expression levels among the HDF cells cultured in three separate bioreactors. (A) The Heatmap shows the expression levels were very similar among the three replicate experiments. Green represents the down-regulated genes and red represents the up-regulated genes. (B) Venn diagram comparing the centralization level of the differentially regulated genes among each of the three replicates. The red represents the up-regulated genes and the green represents the down-regulated genes.

The Ribosome KEGG Pathway included 27 of the SMG sensitive genes; 26 of these genes were down-regulated between 1.5 and 2 fold; only MRPS6 was upregulated (4.5 fold). Interestingly, a number of ribosomal protein genes have also been found to be down-regulated in other studies including our previous study on keratinocytes [21] as well as the NASA Twins Study [16]. The mineral absorption pathway had 10 genes which were all significantly up-regulated in SMG. ATP1A1 and ATP1B3 were both up-regulated by over 2.8 fold. The ferritin genes FTH1 and FTL were up-regulated by 6.6 and 1.6 fold, respectively. Ferritin genes have also been found to be up-regulated as a result of exposure to simulated microgravity in other studies [32]. The metallothionein genes were up-regulated from 7.5 to close to

Pathways	P value	Gene name	FE
hsa03010: Ribosome	1.25×10^{-16}	Ribosomal Protein L7(RPL7), L9 (RPL9), L10A (RPL10A), L11 (RPL11), L12 (RPL12), L18A (RPL18), L27 (RPL27), L27A (RPL27A), L31 (RPL30), L32 (RPL32), L34 (RPL34), L35 (RPL35), L36 (RPL36), L39 (RPL39), S2 (RPS2), S3 (RPS3), S3A (RPS3A), S7 (RPS7), S10 (RPS10), S15A (RPS15A), S17 (RPS17), S18 (RPS18), S19 (RPS19), S29 (RPS29), S23 (RPS23), S24 (RPS24), Mitochondrial Ribosomal ProteinS6 (MRPS6)	7.99
hsa04978: Mineral absorption	9.38×10^{-7}	ATPase Na+/K+ Transporting Subunit Alpha 1 (ATP1A1), ATPase Na+/K+ Transporting Subunit Beta 3 (ATP1B3), Metallothionein 1A (MT1A), 1B (MT1B), 1G MT1G), 1H (MT1H), 1X (MT1X), 2A (MT2A), Ferritin Heavy Chain 1(FTH1), Ferritin Light Chain (FTL)	9.14
hsa04510: Focal adhesion	1.61×10^{-4}	Actin Beta (ACTB), Actin Gamma 1(ACTG1), Caveolin 1 (CAV1), Collagen Type 3 Alpha 1 Chain (COL3A1), Type I Alpha 1 Chain (COL1A1), Type 1 Alpha 2 Chain (COL1A2), Type 6 Alpha 3 Chain (COL6A3) Actinin Alpha 1, Myosin Light Chain 9 (MYL9), 12A (MYL12A), 12B (MYL12B), Calpain 2 (CAPN2), Integrin Beta 1(ITGB1), Filamin A (FLNA),, Vascular endothelial growth factor B (VEGFB), Fibronectin 1(FN1)	3.12
hsa05205: Proteoglycans in cancer	1.30×10^{-3}	ACTB, ACTG1, Cathepsin L (CTSL), Human leukocyte antigen A (HLA-A), B (HLA-B), C (HLA-C), G (HLA-G), Hepatocyte Growth Factor (HGS), Tubulin Beta 6 Class V (TUBB6), Tubulin alpha 1 B (TUBA1B), ITGB1,	2.82
hsa04612: Antigen processing and presentation	2.60×10^{-3}	Beta-2-Microglobulin (B2M), CTSL, HLA-A, HLA-B, HLA-C, HLA-G, Protein Family A (Hsp70) Member 1A Shock Protein (HSPA1A), Protein Family A (Hsp70) Member 8A (HSPA8)	4.23
hsa05012: Parkinson's disease	2.61×10^{-3}	Peptidylprolyl Isomerase F (PPIF), NADH–ubiquinone oxidoreductase chain 1 (ND1), NADH–ubiquinone oxidoreductase chain 3 (ND3) ATP synthase F1 subunit epsilon (ATP5E), ATP Synthase Peripheral Stalk Subunit OSCP (ATP50), NADH:Ubiquinone Oxidoreductase Subunit B4 (NDUFB4), Ubiquinol-Cytochrome C Reductase, Complex III Subunit XI (UQCR11), Cyclooxygenase 1 (COX1) and 2 (COX2), Ubiquitin C-Terminal Hydrolase L1 (UCHL1), Ubiquitin B (UBB)	3.12
hsa05416: Viral myocarditis	2.67×10^{-3}	ACTB, ACTG1, CAV1, HLA-A, HLA-C, HLA-B, HLA-G	4.94
hsa04145: Phagosome	3.88×10^{-3}	ACTB, ACTG1, CTSL, HLA-A, HGS, TUBB6, HLA-C, HLA-B, TUBA1B, ITGB1, HLA-G	2.95
hsa05130: Pathogenic *Escherichia coli* infection	8.15×10^{-3}	ACTB, ACTG1, ARPC2, TUBB6, TUBA1B, ITGB1	4.73
hsa04141: Protein processing in endoplasmic reticulum	8.86×10^{-3}	Heat Shock Protein 90 Beta Family Member 1 (HSP90B1), Valosin Containing Protein (VCP), Defender Against Cell Death 1 (DAD1), DnaJ homolog subfamily A member 1 (DNAJA1), Protein Disulfide Isomerase Family A Member 4 (PDIA4) and 6 (PDIA6), HSPA1A, Heat Shock Protein Family A (Hsp70) Member 5 (HSPA5), CAPN2, HSPA8, Signal Sequence Receptor Subunit 1 (SSR1)	2.62

Pathways	P value	Gene name	FE
hsa00190: Oxidative phosphorylation	1.69×10^{-2}	ND1, ATP5E, NDUFB4, UQCR11, COX2, COX1, ND3, ATP5O, PPA1	2.72
hsa04530: Tight junction	2.03×10^{-2}	ACTB, ACTG1, ACTN1, MYL12B, MYL12A, member RAS oncogene family(RAB13), MYL9	3.24
hsa04512: ECM-receptor interaction	2.03×10^{-2}	CD44, COL3A1, COL6A3, COL1A2, COL1A1, ITGB1, FN1	3.24
hsa04974: Protein digestion and absorption	2.13×10^{-2}	HSP90B1, VCP, DAD1, DNAJA1, PDIA6, HSPA1A, HSPA5, PDIA4, CAPN2, HSPA8, SSR1	3.2
hsa04670: Leukocyte transendothelial migration	2.34×10^{-2}	ACTB, ACTG1, ACTN1, MYL12B, MYL12A, ITGB1, MYL9, THY1	2.8
hsa04260: Cardiac muscle contraction	3.75×10^{-2}	UQCR11, ATP1B3, COX2, COX1, ATP1A1, tropomyosin 2 (beta) (TPM2)	3.22
hsa04611: Platelet activation	4.17×10^{-2}	ACTB, ACTG1, COL3A1, COL1A2, MYL12B, COL1A1, MYL12A, ITGB1	2.48
hsa05100: Bacterial invasion of epithelial cells	4.32×10^{-2}	ACTB, ACTG1, CAV1, ARPC2, ITGB1, FN1	3.09
hsa05332: Graft-versus-host disease	4.71×10^{-2}	HLA-A, HLA-C, HLA-B, HLA-G	4.88
hsa05010: Alzheimer's disease	5.59×10^{-2}	tumor necrosis factor receptor superfamily member 1A(TNFRSF1A), ATP5E, NDUFB4, UQCR11, COX2, COX1, ATP5O, CAPN2, Calmodulin 2 (CALM2)	2.16
hsa05330: Allograft rejection	6.25×10^{-2}	HLA-A, HLA-C, HLA-B, HLA-G	4.35
hsa05131: Shigellosis	7.30×10^{-2}	ACTB, ACTG1, CD44, ARPC2, ITGB1	3.14
hsa04940:Type I diabetes mellitus	8.45×10^{-2}	HLA-A, HLA-C, HLA-B, HLA-G	3.83
hsa04918:Thyroid hormone synthesis	9.43×10^{-2}	HSP90B1, ATP1B3, ATP1A1, HSPA5, PDIA4	2.87

Red indicates that the genes were up-regulated in the data and green indicates the genes were down-regulated in the data. The significance of gene enrichment is evaluated chiefly by the p value with a modified Fisher's exact test. Fold enrichment (FE) is a measure of the magnitude of gene enrichment. P value ≤0.05 and FE ≥1.5 were considered significant and interesting.

Table 1.
KEGG pathway analysis of the 271 differential regulated genes from the HDF cells exposed to 5 days SMG.

12 fold. In a previous study on human keratinocytes in SMG, metallothioneins were also up-regulated [21]. Metallothionein isoforms MT1 and MT2 have been identified as gravity sensitive genes [17, 18, 21, 33–35]. Focal adhesion pathway genes were found to be generally down-regulated by more than 2 fold in the present study. The only exceptions were ITGB1 and VEGFB which were up-regulated by 1.5 and 1.8 fold, respectively. There were 14 genes in the proteoglycans in cancer category, with 9 genes being up-regulated and 4 down-regulated; these included several major gravity sensitive genes identified previously, CAV1, FN1, DCN, and CD44 [17]. In the antigen processing and presentation pathway, the 8 genes represented were all up-regulated. These include the HLA genes, the closely related B2M, as well as CTSL, HSPA1A and HSPA8. In a previous study, HLA-G is also up-regulated in the SMG treated human keratinocytes [21]. In normal gravity environment, HLA-G has been shown to have direct inhibitory effect on T, APC, and NK cell functions and induces suppressor T-cells [36]. It has been found to be present in neurological

disorders [37]. HLA-G is considered a stress inducible gene [38]. HLA-G also plays a role in tumor-driven immune escape mechanism of cancer cells during the later phase in host and tumor cell interactions. When HLA-G is expressed, it can result in an immune suppressive function [38–40]. HLA-G expression has been found to correlate with low frequency of rejection in some forms of organ transplants [39]. Heat shock proteins HSPA1A and HSPA8 have been identified as being gravity sensitive in several studies. In some of these studies, HSPA1A has been shown to be up-regulated [21, 32, 41]. They were also shown to be down-regulated [42].

3.3 Validation of microarray results through northern blotting analysis

The quality of a gene list from a high throughput study is essential for a successful functional analysis in DAVID [29]. The high throughput microarray data of HDF in SMG presented above was further validated by performing Northern blot analysis.

Northern blotting measures the abundance as well as the size of the RNA of interest [21]. In agreement with the microarray data, the Northern results showed

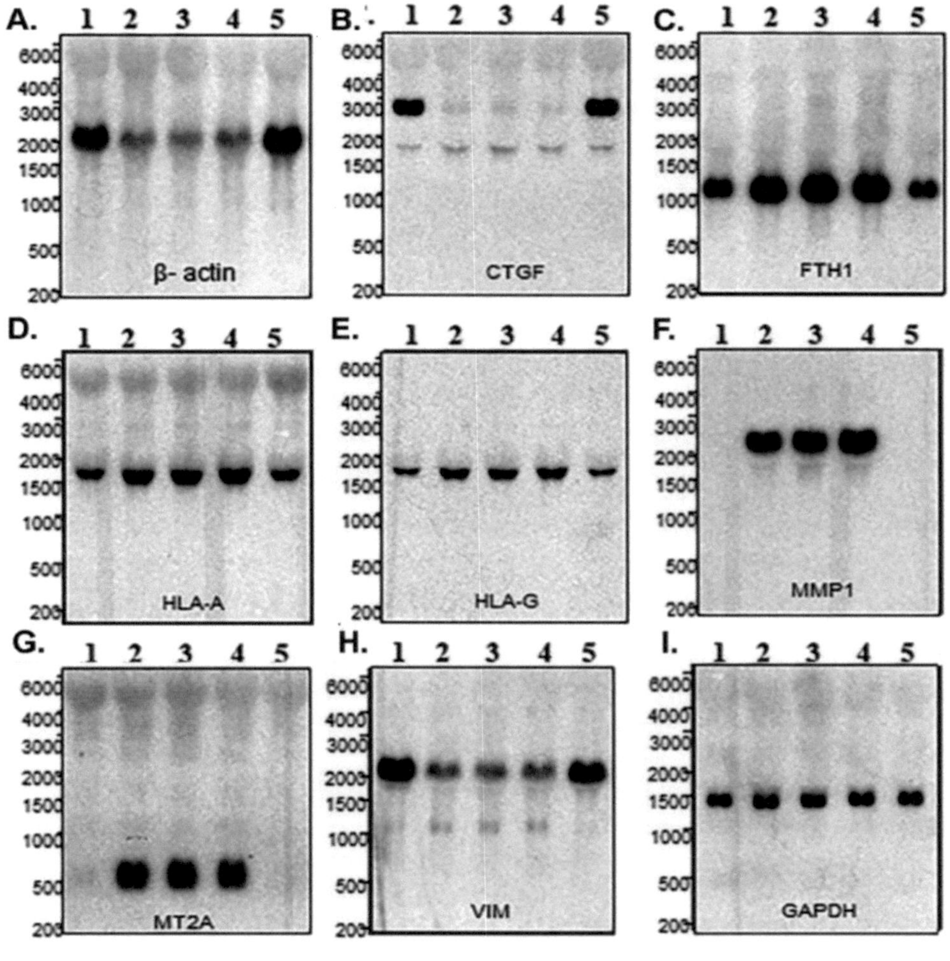

Figure 3.
Microarray data validation using northern blot analysis. 10 μg of total RNA of the fibroblast cells from each of the three RWV bioreactors (3-D spheres) were loaded in lanes 2–4. Lanes 1 and 5 are control RNA samples from the cells grown at normal gravity. The northern blot was probed and striped repeatedly using cDNA probe fragments from genes indicated at the bottom of each panel (A–I). RNA size markers in nucleotides were labeled along the left side of each panel.

that the mRNA levels of β-actin, CTGF, and VIM were down regulated, while that of FTH1, HLA-A, HLA-B (data not shown), HLA-G, MMP1, and MT2A were upregulated (**Figure 3**). In addition, Northern analysis showed that the expression level of glyceraldehyde-3-phosphate dehydrogenase (GAPDH) was consistent in the presence and absence of microgravity (**Figure 3I**).

3.4 Comparison of HDF in SMG gene expression data with that from the NASA twins study

The NASA Twins study is very substantial in undertaking with extensive data sets [14]. It was very appealing to compare the current data of HDF in SMG with the recently published gene expression data from the NASA Twins study. Since a large amount of data is available, it is more manageable to first focus on the gene expression data from the samples taken from inflight first half (up to 6 months) and the second half (up to 1 year), generated from the inflight astronaut's PBMCs. Such comparison of the two gene expression data sets gave 194 overlapping differentially expressed genes, or about 72%, of the genes found to be differentially regulated in the current study with HDF in SMG were also differentially regulated in the selected data set from the NASA study. Of these 194 genes, 86 had a similar expression pattern (regulated in the same direction) to the HDF in SMG data and 108 had the opposite expression pattern. A Heatmap was generated as a way of better visualizing the similarities between the data points (**Figure 4**).

3.5 KEGG pathways of the overlapping genes

As a way of comparing and better understanding the potential relationships between the HDF SMG gene expression data and the NASA Twin gene expression data from **Table 2** [16], we uploaded the gene list of the genes that showed same direction expression patterns (**Table 2**) and the genes that showed opposite expression patterns (**Table 3**) to DAVID in order to generate KEGG Pathway information.

After processing the list of genes with the same expression patterns through DAVID, a total of 7 KEGG pathways were statistically significant ($P \leq 0.05$) Interestingly, the genes enriched in all 9 KEGG pathways had >3.5 fold enrichment (**Table 2**). All the genes represented in the KEGG pathways were down-regulated in both sets of data. The Ribosome pathway had the greatest number of genes represented. The down-regulation of the ribosomal protein genes is consistent with our previous study in keratinocytes in SMG [21].

KEGG pathway analysis of the list of genes with opposite direction of expression regulation produced a total of 15 KEGG Pathways that were statistically significant ($P \leq 0.05$). Interestingly, the genes in all the 21 KEGG pathways had >2.5 fold enrichment (**Table 3**).

The protein processing in endoplasmic reticulum (ER) pathway had the greatest number of genes represented. In the current study of HDF in SMG, the genes represented in the ER pathway were all up-regulated whereas they were down-regulated in the NASA Twin Study data [16]. Heat shock proteins HSPA1A and HSPA8 have been identified as being gravity sensitive in several studies. In some of these studies, HSPA1A has been shown to be up-regulated [21, 32, 41] and in one study in addition to the NASA Twin Study was shown to be down-regulated [42]. Collagens also seem to be mixed in differential regulation in microgravity. In our current study, COL1, 2, and 6 were down-regulated but were up-regulated in the Twin study. In some studies COL1 has been shown to be up-regulated [43] while in others it has been shown to be down-regulated [33, 44]. FN1, which we identified as a putative

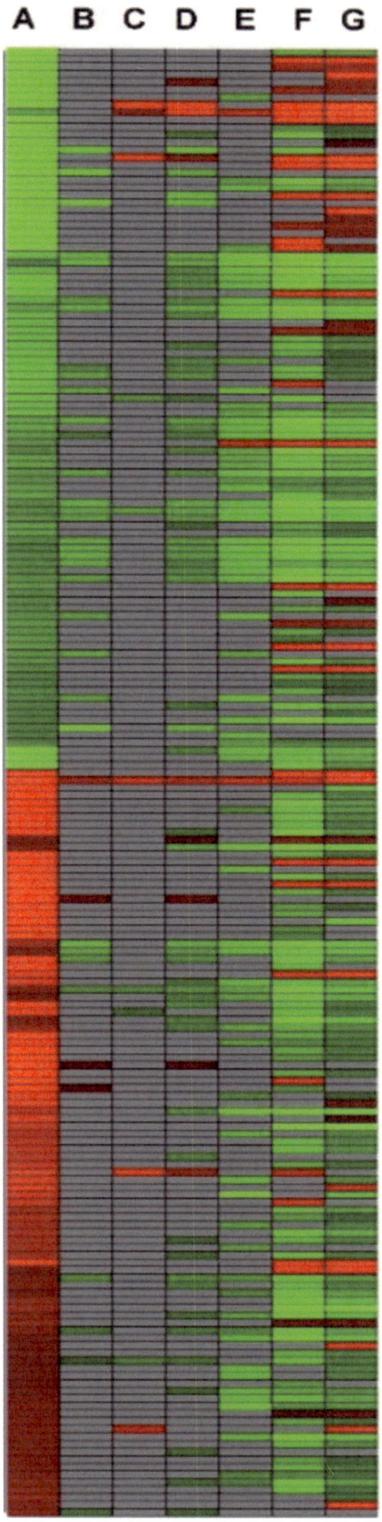

Figure 4.
Heatmap comparison of HDF in SMG gene expression data with that from the NASA twins study Heatmap is used to compare the HDF SMG data (column A) with the inflight, first half of the twins study ribodepleted (column B), PolyA+ (column C), multivariant (column D), and the inflight, second half of the twins study ribodepleted (column E), PolyA+ (column F), and multivariant (column G). Green were down-regulated genes, red represented up-regulated genes, and gray indicates the presence of no corresponding data.

Pathways	P Value	Gene name	FE
hsa03010:Ribosome	3.93×10^{-27}	RPL35, RPL27A, RPL36, RPS15A, RPS2, RPL39, RPS3, RPL32, RPL7, RPL31, RPS3A, RPL9, RPL34, RPL11, RPL10A, RPL12, RPS23, RPS24, RPL27, RPS7, RPS18, RPS19, RPL18A, RPS17, RPS10	22.18
hsa04530:Tight junction	5.32×10^{-3}	ACTB, ACTG1, MYL12B, MYL12A, MYL9	6.936
hsa05130:Pathogenic Escherichia coli infection	8.07×10^{-3}	ACTB, ACTG1, ARPC2, TUBA1B	9.465
hsa04670:Leukocyte transendothelial migration	1.40×10^{-2}	ACTB, ACTG1, MYL12B, MYL12A, MYL9	5.247
hsa04510:Focal adhesion	2.55×10^{-2}	ACTB, ACTG1, MYL12B, MYL12A, CAPN2, MYL9	3.515
hsa04810:Regulation of actin cytoskeleton	2.74×10^{-2}	ACTB, ACTG1, ARPC2, MYL12B, MYL12A, MYL9	3.448
hsa04921:Oxytocin signaling pathway	3.32×10^{-2}	ACTB, MYL6, ACTG1, CALM2, MYL9	4.023
hsa04611:Platelet activation	8.88×10^{-2}	ACTB, ACTG1, MYL12B, MYL12A	3.713
hsa05131:Shigellosis	9.53×10^{-2}	ACTB, ACTG1, ARPC2	5.657

The gene symbols in green color indicated down-regulation. The significance of gene enrichment is evaluated chiefly by the p value with a modified Fisher's exact test. FE is a measure of the magnitude of gene enrichment. P value ≤ 0.05 and FE ≥ 1.5 were considered significant and interesting, respectively.

Table 2.
KEGG pathways generated from comparison of genes differentially regulated in the same direction between the HDF and NASA twins data.

Pathways	P value	Gene name	FE
hsa04141:Protein processing in endoplasmic reticulum	2.93×10^{-5}	HSP90B1, VCP, DAD1, DNAJA1, PDIA6, HSPA1A, HSPA5, PDIA4, HSPA8, SSR1	5.81
hsa04512:ECM-receptor interaction	1.70×10^{-4}	CD44, COL3A1, COL6A3, COL1A2, COL1A1, ITGB1, FN1	7.91
hsa04612:Antigen processing and presentation	7.62×10^{-4}	HLA-A, HLA-C, HSPA1A, HLA-B, HSPA8, B2M	7.76
hsa04978:Mineral absorption	7.86×10^{-4}	ATP1B3, ATP1A1, MT1X, FTH1, FTL	11.2
hsa04974:Protein digestion and absorption	1.48×10^{-3}	ATP1B3, COL3A1, COL6A3, COL1A2, ATP1A1, COL1A1	6.7
hsa04918:Thyroid hormone synthesis	4.41×10^{-3}	HSP90B1, ATP1B3, ATP1A1, HSPA5, PDIA4	9.1
hsa04510:Focal adhesion	1.37×10^{-2}	CAV1, COL3A1, COL6A3, COL1A2, COL1A1, ITGB1, FN1	7.02
hsa05134:Legionellosis	1.48×10^{-2}	ARF1, VCP, HSPA1A, HSPA8	3.44
hsa04151:PI3K-Akt signaling pathway	1.67×10^{-2}	HSP90B1, COL3A1, COL6A3, COL1A2, YWHAQ, COL1A1, ITGB1, FN1, DDIT4	3.34
hsa05416:Viral myocarditis	1.71×10^{-2}	CAV1, HLA-A, HLA-C, HLA-B	6.9
hsa04144:Endocytosis	2.74×10^{-2}	CAV1, ARF1, HLA-A, HLA-C, HSPA1A, HLA-B, HSPA8	2.56

Pathways	P value	Gene name	FE
hsa05169:Epstein–Barr virus infection	2.92×10^{-2}	CD44, VIM, HLA-A, HLA-C, HLA-B	4.64
hsa05332:Graft-versus-host disease	3.96×10^{-2}	HLA-A, HLA-C, HLA-B	4.06
hsa05205:Proteoglycans in cancer	4.23×10^{-2}	CAV1, CD44, CD63, DDX5, ITGB1, FN1	2.85
hsa05330:Allograft rejection	4.87×10^{-2}	HLA-A, HLA-C, HLA-B	8.93
hsa04514:Cell adhesion molecules (CAMs)	5.37×10^{-2}	HLA-A, HLA-C, HLA-B, ITGB1	7.97
hsa04145:Phagosome	6.32×10^{-2}	HLA-A, HLA-C, HLA-B, ITGB1	3.46
hsa04940:Type I diabetes mellitus	6.60×10^{-2}	HLA-A, HLA-C, HLA-B	3.28
hsa03050:Proteasome	7.16×10^{-2}	PSMB4, PSMB7, PSMA6	7.02
hsa05320:Autoimmune thyroid disease	9.55×10^{-2}	HLA-A, HLA-C, HLA-B	6.7
hsa05145:Toxoplasmosis	9.79×10^{-2}	PPIF, HSPA1A, ITGB1, HSPA8	5.67

The green indicates genes that were down-regulated and the red indicates genes that were up regulated in the HDF data set. The significance of gene enrichment is evaluated chiefly by the p value with a modified Fisher's exact test. FE is a measure of the magnitude of gene enrichment. P value ≤0.05 and FE ≥1.5 were considered significant and interesting, respectively.

Table 3.
KEGG pathways generated from a comparison of genes differentially regulated in the opposite direction between the HDF and NASA twins data.

"space gene" [17, 18] has also been shown to have variations in expression patterns. It has been shown to be down-regulated in several studies [34, 44, 45] and up-regulated in others [16, 43, 46].

4. Discussion

In the current study, the gene expression profile of HDF grown in 5 day SMG was first displayed and validated (**Figures 1–3**, **Table 1**). The high throughput cDNA microarray data of HDF in 5 day SMG was then used to compare with the high throughput RNA sequencing data from an astronaut's PBMCs during a long term inflight ISS (**Figure 4**, **Tables 2** and **3**). Amazingly, about 72% of the 271 microgravity sensitive genes of HDF in SMG, were also differentially regulated in the NASA Twins 6- and 12-month inflight data. These 194 overlapping genes were identified as putative microgravity sensitive space genes, because the human dermal fibroblast cell line was only exposed to SMG, no radiation nor other space related environmental factor was involved. However, other factors such as cell type difference and space radiation, may also influence the expression of microgravity sensitive genes. Indeed, among these microgravity sensitive genes, 86 genes showed the same expression pattern in both simulated and real microgravity conditions while the other 108 genes displayed opposite direction of regulation.

It is remarkable that as many as 86 genes were found to have the same directions of expression regulation between very different settings of studies. It is most likely that these genes were the main players in cellular response to microgravity environment. When the 86 microgravity sensitive genes with the same expression regulation trends were subjected to KEGG pathway analysis, they were represented in seven significant pathways where they were all downregulated (**Table 2**). Most notably, both sets of global gene expression data showed the down regulation of 25 ribosomal protein genes. The genes in pathogenic Escherichia coli infection pathway and leukocyte trans-endothelial migration pathway were all down-regulated in microgravity (**Table 2**), which may contribute to the decreased immune resistant to microbial infection in spaceflight. In addition, genes in the cytoskeleton network (ACTB, ACTG1, ARPC2, MYL12B, MYL12A, MYL9), focal adhesion (ACTB, ACTG1, MYL12B, MYL12A, CAPN2, MYL9), as well as tight junction (ACTB, ACTG1, MYL12B, MYL12A, MYL9) pathways were also downregulated in both sets of data. In combination with the downregulation of extracellular matrix proteins (**Table 1**) such as COL1A1, COL1A2, COL3A1 in HDF, the data indicated an overall decrease in bone matrix and skeletal muscle synthesis and increased catabolism (e.g. MMP1 increased sharply). Furthermore, genes in the oxytocin signaling pathway (ACTB, MYL6, ACTG1, CALM2, MYL9), which is involved in smooth muscle contraction and stress management, were also down in their expression levels in both microgravity data sets. Malfunction of this pathway has been implicated in depression, autism, and schizophrenia [47]. Overall, the data in **Table 2** gave a strong mechanistic connection to the main symptoms, such as skin problems and immunological stress, vascular changes, muscle atrophy and bone density alteration that were associated with the weightlessness of space flight. Indeed, the altered expression of these 86 microgravity sensitive genes affected many fundamental molecular functions (data not shown), including structural constituent of ribosome, RNA binding, protein binding, metal ion binding, structural constituent of the cytoskeleton, cadherin binding involved in cell-cell adhesion, extracellular matrix binding, etc. Many biological processes in these cells (data not shown), such as SRP-dependent co-translational protein targeting to membrane, translation initiation and elongation, mRNA stability, muscle contraction, regulation of cell shape, among others, were also significantly impacted. These genes with common trend of expression regulation in microgravity would most likely expand the list of putative major space genes and microgravity sensitive pathways [17, 18]. A substantial amount of information was derived from the current work which may necessitate more detailed analysis and discussion in future communications.

The number of overlapping microgravity sensitive genes was substantial considering the many differences between the two study settings. The result from this comparative analysis further validated the effectiveness of the bioreactors for SMG cell culture. The identification of 86 genes (**Table 2**) with the same direction of regulation in two different study settings is very substantial and unique. These genes should be considered best candidates for major microgravity sensitive genes because one of the two studies, the current study, only involves simulated microgravity, while the other study, the NASA Twins Study involves true spaceflight environment with microgravity and space radiation. This comparative analysis here enabled the differentiation of the microgravity effect alone on the differentially expressed genes from the human astronaut spaceflight gene expression data. The identification of the overlapping significant genes regulated in the opposite direction rendered important insight into human gene activity changes in very different study systems. The HDF cell line in SMG versus the human astronaut in ISS, adjusted their expressions toward adaptation to the simple SMG as well as the true space environment of both space microgravity and space radiation. The 108 microgravity sensitive genes with opposite directions of expression regulation could also be of major significance in the microgravity adaptation process. Compared to the single cell line in SMG alone for the HDF cells, the cell samples from the inflight astronaut was exposed to various other factors such as space radiation, in addition to the microgravity of the ISS. The more complex space environment may require the significant genes to modify their expression toward adaptation. Expression patterns in this group of genes could provide insight into our understanding regarding the interplay among different cellular gene functions in human adaptation to microgravity and space radiation (**Table 3**).

Our previous studies on gene expression in HEK cells grown in SMG has 43 genes overlapping with the HDF data in the current communication; of which, 23 genes were regulated in the same direction and 20 were regulated in the opposite direction [21]. These two different types of cells require different culture conditions and perform different roles in the skin. It is understandable that they have their characteristic expression profiles in response to the simulated microgravity environment. However, the number of overlapping genes were also substantial. In a previous review paper comparing various microarray based gene expression studies on microgravity effects, an initial list of 129 genes were identified as putative microgravity sensitive genes or major space genes [17]. In the current study, 12 out of the 194 genes that were significantly differentially regulated in both the HDF cells and the PBMCs, are also in the group of the putative major spaces genes [17], with 4 genes regulated in the same direction (MMP1, GPNMB, RPL10A, and ANXA2) and 8 genes (CAV1, CD44, CD59, CYR61, FN1, HSPA1A, MT1X, and PDIA4) changed their expression in the opposite directions in response to microgravity. Continued microgravity research in space and the readily controlled simulated microgravity bioreactors would provide valuable information toward the identification of major gravity sensitive genes, or simply, the major space genes. With more data available, the molecular and cellular mechanisms underlying the microgravity response could be better understood. Elucidation the molecular mechanism of human space adaptation response is an important aspect toward safer space experience and human health in general. It is evident that continued microgravity research is beneficial to healthy living in space and on earth.

5. Conclusion

The identification of 271 genes of HDF significantly differentially regulated by SMG provided a set of data for more detailed mechanistic studies; 72% of these microgravity sensitive genes were also reported in the high throughput gene expression data in the recent NASA Twins' Study.

The identification of the large number of overlapping genes between the HDF in SMG and astronaut's PBMCs indicates microgravity alone, without space radiation, was able to elicit an adaptive response involving a set of about 200 genes.

Acknowledgements

This work was partially supported by NASA JSC Grant No. NCC9-165.

Abbreviations/glossary

ACTB	actin beta
ACTG1	actin gamma 1
ANXA2	annexin A2
ATP1A1	ATPase Na+/K+ transporting subunit alpha 1
ATP1B3	ATPase Na+/K+ transporting subunit beta 3
ATP50	ATP synthase peripheral stalk subunit OSCP
ATP5E	ATP synthase F1 subunit epsilon
B2M	beta-2microglobulin
CALM2	calmodulin 2
CAPN2	calpain 2

CARMAweb	comprehensive R based microarray analysis web frontend
COL	collagen
COX	cyclooxygenase
CTGF	connective tissue growth factor
CTSL	cathepsin L
CYR61	cysteine rich angiogenic inducer 61
DAD	defender against cell death 1
DAVID	database for annotation, visualization and integrated discovery
DNAJA1	DnaJ homolog subfamily A member 1
FLNA	filamin A
FN1	fibronectin
FTH1	ferritin heavy chain 1
FTL	ferritin light chain
GAPDH	glyceraldehyde-3-phosphate dehydrogenase
GO	gene ontology
GPNMB	glycoprotein NMB
HARVs	high aspect ratio vessels
HDF	human dermal fibroblast
HLA	human leukocyte antigen
HSP	heat shock protein
ISS	international space station
ITGB1	integrin beta 1
KEGG	kyoto encyclopedia of genes and genomes
MMP1	matrix metallopeptidase 1
MRPS6	mitochondrial ribosomal protein 6
MT	metallothionein
MYL	myosin light chain
ND	NADH–ubiquinone oxidoreductase chain
NDUFB4	NADH:ubiquinone oxidoreductase subunit B4
PBMCs	peripheral blood mononuclear cells
PDIA	Protein disulfide isomerase family A member
PPIF	peptidylprolyl isomerase F
RAB13	member RAS oncogene family
RCCS	rotary cell culture system
RP	ribosomal protein
RWV	rotating wall vessel
SMG	simulated microgravity
SSR1	signal sequence receptor subunit 1
TNFRSF1A	tumor necrosis factor receptor superfamily member 1A
TPM2	tropomyosin beta chain
TUBA1B	tubulin alpha 1 B
TUBB6	tubulin beta 6 class V
UBB	ubiquitin B
UCHL1	ubiquitin C-terminal hydrolase L1
UQCR11	ubiquinol-cytochrome C reductase, complex III subunit XI
VCP	valosin containing protein
VEGFB	vascular endothelial growth factor B

Author details

Jade Q. Clement
Department of Chemistry, Texas Southern University, Houston, Texas, United States

*Address all correspondence to: jade.clement@tsu.edu

References

[1] Platts SH, Bairey Merz CN, Barr Y, Fu Q, Gulati M, Hughson R, et al. Effects of sex and gender on adaptation to space: Cardiovascular alterations. Journal of Women's Health. 2014;**23**(11):950-955. DOI: 10.1089/jwh.2014.4912

[2] Fritsch-Yelle JM, Charles JB, Jones MM, Wood ML. Microgravity decreases heart rate and arterial pressure in humans. Journal of Applied Physiology. 1996;**80**(3):910-914. DOI: 10.1152/jappl.1996.80.3.910

[3] Leach CS, Dietlein LF, Pool SL, Nicogossian AE. Medical considerations for extending human presence in space. Acta Astronautica. 1990;**21**(9):659-666. DOI: 10.1016/0094-5765(90)90077-X

[4] Cogoli A. Space flight and the immune system. Vaccine. 1993;**11**(5):496-503. DOI: 10.1016/0264-410X(93)90217-L

[5] Arfat Y, Xiao WZ, Iftikhar S, Zhao F, Li DJ, Sun YL, et al. Physiological effects of microgravity on bone cells. Calcified Tissue International. 2014;**94**(6):569-579. DOI: 10.1007/s00223-014-9851-x

[6] Gerbaix M, Gnyubkin V, Farlay D, Olivier C, Ammann P, Courbon G, et al. One-month spaceflight compromises the bone microstructure, tissue-level mechanical properties, osteocyte survival and lacunae volume in mature mice skeletons. Scientific Reports. 2017;**7**(1):2659. DOI: 10.1038/s41598-017-03014-2

[7] Hayes J. The first decade of ISS exercise: Lessons learned on expeditions 1-25. Aerospace Medicine and Human Performance. 2015;**86**(Suppl 12):A1-A6. DOI: 10.3357/AMHP.EC01.2015

[8] Aubert AE, Beckers F, Verheyden B. Cardiovascular function and basics of physiology in microgravity. Acta Cardiologica. 2005;**60**(2):129-151. DOI: 10.2143/AC.60.2.2005024

[9] Harding CP, Vargis E. Muscle atrophy marker expression differs between rotary cell culture system and animal studies. BioMed Research International. 2019;**2019**:2042808. DOI: 10.1155/2019/2042808

[10] Trappe S, Costill D, Gallagher P, Creer A, Peters JR, Evans H, et al. Exercise in space: Human skeletal muscle after 6 months aboard the International Space Station. Journal of Applied Physiology. 2009;**106**(4):1159-1168. DOI: 10.1152/japplphysiol.91578.2008

[11] Sonnenfeld G. Use of animal models for space flight physiology studies, with special focus on the immune system. Gravitational and Space Biology Bulletin. 2005;**18**(2):31-35. Available from: http://gravitationalandspacebiology.org/index.php/journal/article/view/344/345

[12] Crucian B, Babiak-Vazquez A, Johnston S, Pierson DL, Ott CM, Sams C. Incidence of clinical symptoms during long-duration orbital spaceflight. International Journal of General Medicine. 2016;**9**:383-391. DOI: 10.2147/IJGM.S114188

[13] Tauber S, Lauber BA, Paulsen K, Layer LE, Lehmann M, Hauschild S, et al. Cytoskeletal stability and metabolic alterations in primary human macrophages in long-term microgravity. PLoS One. 2017;**12**(4):e0175599. DOI: 10.1371/journal.pone.0175599

[14] Abdallah F, Mijouin L, Pichon C. Skin immune landscape: Inside and outside the organism. Mediators of Inflammation. 2017;**2017**:5095293. DOI: 10.1155/2017/5095293

[15] Kuhbacher A, Henkel H, Stevens P, Grumaz C, Finkelmeier D, Burger-Kentischer A, et al. Central role for dermal fibroblasts in skin model protection against Candida albicans.

The Journal of Infectious Diseases. 2017;**215**(11):1742-1752. DOI: 10.1093/infdis/jix153

[16] Garrett-Bakelman FE, Darshi M, Green SJ, Gur RC, Lin L, Macias BR, et al. The NASA twins study: A multidimensional analysis of a year-long human spaceflight. Science. 2019;**364**(6436). DOI: 10.1126/science.aau8650

[17] Abdelmoaty H, Hammond TG, Wilson BL, Birdsall HH, Clement JQ. dentification of putative major space genes using genome-wide literature data. In: Ekinci D, editor. Biotechnology. London: InTech; 2015. pp. 93-116. DOI: 10.5772/60412

[18] Clement JQ. Gene expression microarrays in microgravity research: Toward the identification of major space genes. In: Agbo EC, editor. Innovations in Biotechnology. InTech; 2012. pp. 319-348. DOI: 10.5772/28693

[19] Bauer J, Bussen M, Wise P, Wehland M, Schneider S, Grimm D. Searching the literature for proteins facilitates the identification of biological processes, if advanced methods of analysis are linked: A case study on microgravity-caused changes in cells. Expert Review of Proteomics. 2016;**13**(7):697-705. DOI: 10.1080/14789450.2016.1197775

[20] Clement JQ. Microarray profiling of genome-wide expression regulation in response to environmental exposures. In: Fung GPC, editor. A Practical Guide to Bioinformatics Analysis. Brisbane: Iconcept Press; 2010. pp. 23-40. Available from: https://www.inconcept.com/book/a-practical-guide-to-bioinformatics-analysis/1100012/100900085/download/1009000085.pdf

[21] Clement JQ, Lacy SM, Wilson BL. Gene expression profiling of human epidermal keratinocytes in simulated microgravity and recovery cultures. Genomics, Proteomics & Bioinformatics. 2008;**6**(1):8-28. DOI: 10.1016/S1672-0229(08)60017-0

[22] Clement JQ, Qian L, Kaplinsky N, Wilkinson MF. The stability and fate of a spliced intron from vertebrate cells. RNA. 1999;**5**(2):206-220. DOI: 10.1017/s1355838299981190

[23] Wilkinson MF. Purification of RNA. In: Brown TA, editor. Essential Molecular Biology: A Practical Approach. 1. New York: Oxford University Press; 1991. pp. 69-86. DOI: 10.1016/0962-8924(92)90251-H

[24] Wolber PK, Collins PJ, Lucas AB, De Witte A, Shannon KW. The Agilent in situ-synthesized microarray platform. Methods in Enzymology. 2006;**410**:28-57. DOI: 10.1016/S0076-6879(06)10002-6

[25] Rainer J, Sanchez-Cabo F, Stocker G, Sturn A, Trajanoski Z. CARMAweb: Comprehensive R- and bioconductor-based web service for microarray data analysis. Nucleic Acids Research. 2006;**34**:W498-W503. DOI: 10.1093/nar/gkl038

[26] Benjamini Y, Hochberg Y. Controlling the false discovery rate: A practical and powerful approach to multiple testing. Journal of the Royal Statistical Society, Series B. 1995;**57**(1):289-300. Available from: http://www.math.tau.ac.il/~ybenja/MyPapers/benjamini_hochberg1995.pdf

[27] Kanehisa M, Goto S. KEGG: Kyoto encyclopedia of genes and genomes. Nucleic Acids Research. 2000;**28**(1):27-30. DOI: 10.1093/nar/28.1.27

[28] Kanehisa M, Sato Y, Furumichi M, Morishima K, Tanabe M. New approach for understanding genome variations in KEGG. Nucleic Acids Research. 2019;**47**(D1):D590-D595. DOI: 10.1093/nar/gky962

[29] Huang d W, Sherman BT, Lempicki RA. Systematic and integrative analysis of large gene lists using DAVID bioinformatics resources. Nature Protocols. 2009;**4**(1):44-57. DOI: 10.1038/nprot.2008.211

[30] Sturn A, Quackenbush J, Trajanoski Z. Genesis: Cluster analysis of microarray data. Bioinformatics. 2002;**18**(1):207-208. DOI: 10.1093/bioinformatics/18.1.207

[31] Breitkreutz BJ, Jorgensen P, Breitkreutz A, Tyers M. AFM 4.0: A toolbox for DNA microarray analysis. Genome Biology. 2001;**2**(8): SOFTWARE0001. DOI: 10.1186/gb-2001-2-8-software0001

[32] Grosse J, Wehland M, Pietsch J, Schulz H, Saar K, Hubner N, et al. Gravity-sensitive signaling drives 3-dimensional formation of multicellular thyroid cancer spheroids. The FASEB Journal. 2012;**26**(12): 5124-5140. DOI: 10.1096/fj.12-215749

[33] Allen DL, Bandstra ER, Harrison BC, Thorng S, Stodieck LS, Kostenuik PJ, et al. Effects of spaceflight on murine skeletal muscle gene expression. Journal of Applied Physiology. 2009;**106**(2):582-595. DOI: 10.1152/japplphysiol.90780.2008

[34] Nikawa T, Ishidoh K, Hirasaka K, Ishihara I, Ikemoto M, Kano M, et al. Skeletal muscle gene expression in space-flown rats. The FASEB Journal. 2004;**18**(3):522-524. DOI: 10.1096/fj.03-0419fje

[35] Hammond TG, Benes E, O'Reilly KC, Wolf DA, Linnehan RM, Taher A, et al. Mechanical culture conditions effect gene expression: Gravity-induced changes on the space shuttle. Physiological Genomics. 2000;**3**(3):163-173. DOI: 10.1152/physiolgenomics.2000.3.3.163

[36] Mouillot G, Marcou C, Zidi I, Guillard C, Sangrouber D, Carosella ED, et al. Hypoxia modulates HLA-G gene expression in tumor cells. Human Immunology. 2007;**68**(4):277-285. DOI: 10.1016/j.humimm.2006.10.016

[37] Wiendl H. HLA-G in the nervous system. Human Immunology. 2007;**68**(4):286-293. DOI: 10.1016/j.humimm.2006.10.021

[38] Alegre E, Rizzo R, Bortolotti D, Fernandez-Landazuri S, Fainardi E, Gonzalez A. Some basic aspects of HLA-G biology. Journal of Immunology Research. 2014;**2014**:657625. DOI: 10.1155/2014/657625

[39] Loustau M, Wiendl H, Ferrone S, Carosella ED. HLA-G 2012 conference: The 15-year milestone update. Tissue Antigens. 2013;**81**(3):127-136. DOI: 10.1111/tan.12053

[40] Donadi EA, Castelli EC, Arnaiz-Villena A, Roger M, Rey D, Moreau P. Implications of the polymorphism of HLA-G on its function, regulation, evolution and disease association. Cellular and Molecular Life Sciences. 2011;**68**(3):369-395. DOI: 10.1007/s00018-010-0580-7

[41] Ward NE, Pellis NR, Risin SA, Risin D. Gene expression alterations in activated human T-cells induced by modeled microgravity. Journal of Cellular Biochemistry. 2006;**99**(4): 1187-1202. DOI: 10.1002/jcb.20988

[42] Chang TT, Hughes-Fulford M. Molecular mechanisms underlying the enhanced functions of three-dimensional hepatocyte aggregates. Biomaterials. 2014;**35**(7):2162-2171. DOI: 10.1016/j.biomaterials.2013.11.063

[43] Chopard A, Hillock S, Jasmin BJ. Molecular events and signalling pathways involved in skeletal muscle disuse-induced atrophy and the impact of countermeasures. Journal of Cellular and Molecular Medicine.

2009;**13**(9B):3032-3050. DOI: 10.1111/j.1582-4934.2009.00864.x

[44] Sheyn D, Pelled G, Netanely D, Domany E, Gazit D. The effect of simulated microgravity on human mesenchymal stem cells cultured in an osteogenic differentiation system: A bioinformatics study. Tissue Engineering. Part A. 2010;**16**(11):3403-3412. DOI: 10.1089/ten.tea.2009.0834

[45] Dapp C, Schmutz S, Hoppeler H, Fluck M. Transcriptional reprogramming and ultrastructure during atrophy and recovery of mouse soleus muscle. Physiological Genomics. 2004;**20**(1):97-107. DOI: 10.1152/physiolgenomics.00100.2004

[46] Qian A, Di S, Gao X, Zhang W, Tian Z, Li J, et al. cDNA microarray reveals the alterations of cytoskeleton-related genes in osteoblast under high magneto-gravitational environment. Acta Biochimica et Biophysica Sinica. 2009;**41**(7):561-577. DOI: 10.1093/abbs/gmp041

[47] Shen H. Neuroscience: The hard science of oxytocin. Nature. 2015;**522**(7557):410-412. DOI: 10.1038/522410a

Chapter 9

Environmental Factors Affecting the Expression of Bilateral-Symmetrical Traits in Plants

Sergey Baranov, Igor Vinokurov and Lubov Fedorova

Abstract

In recent years, there has been a growing interest in the problem of asymmetry of bilateral traits in plants. Three types of bilateral asymmetry are found in the leaf blade, of interest to ecologists and evolutionists. A brief review of the methods used in testing bilateral asymmetry and developmental stability discusses their role in the development of homeostasis and ontogenesis. Intra- and interspecific differences are considered on the example of woody plants under the influence of factors influencing the expression of bilaterally symmetry. The influence of stress on the manifestation of asymmetric traits is considered. Apparently, the climate and topography of the area play a more important role, determining the plastic and fluctuating variability. The relationship of plasticity, evolutionary canalization, and development stability is considered on the example of woody and cultivated plants. Plasticity and fluctuation variability are in a relationship coordinated by climatic conditions, primarily lighting and temperature. This, in turn, determines the mechanisms of gene regulatory networks. Thus, phenogenetics, which studies the patterns and mechanisms of gene expression and ontogenesis, is based on the data from field botanical studies of plant shape and asymmetry. Epigenetic and population studies of phenotypic variations play a role in standardizing and finding suitable plant species and varieties.

Keywords: bilateral asymmetry, fluctuating variability, gene regulatory networks

1. Three types of bilateral asymmetry

One of the promising areas of monitoring for the environment is bioindication by determining the developmental stability (DS) of plants, including woody ones. Database on the developmental stability of different species of plants is to be complementing other data sets, such as chemical contamination of air, soil, and water.

Fluctuating asymmetry (FA) is a kind of asymmetry used to assess the stability of development, as the organism's ability to regulate its development on the phylogenetic level.

The concept of developmental noise was introduced by Worthington [1], developed and completed in the works of foreign and Russian scientists at the end of the 20th century [2–6]. This term originally meant the factors that lead to deviations from bilateral symmetry, such as metabolic rate, the concentration of biochemical regulator molecules, diffusion of substances, temperature gradient, growth, and cell death.

Fluctuating asymmetry is most often referred to as the minor variation of morphological characters from perfect symmetry. FA is an expression of ontogenetic noise and, hence, the level of developmental stability. The value of FA is defined as a nondirectional minor deviation from the strict bilateral symmetry. At the same time, the value of fluctuating asymmetry means independence in phenotypic realization on the left and right lateral sides. Thus, currently FA is the only characteristic that allows studying the developmental stability in a morphogenetic aspect. Fluctuating asymmetry is a common phenomenon among the flora of higher plants (Embryophyta). FA is most frequently expressed in the leaf blade, since they are commonly bilaterally asymmetric. As can be seen from many sources, the large number of plastic (metric) traits and meristic (countable) traits promotes the robust FA testing. Integral index is an average means of FA some traits. This takes into account only the trait values that are not correlated with each other [7, 8].

Deviations from genetically predetermined rules are presented as a fact of random expression genotype under the influence of environmental impact. Based on the genetically determined reaction, fluctuating asymmetry is also considered as a deviation from the reaction norms. The fluctuating asymmetry has to be distinguished from phenodeviation having more genetic sense. Phenodeviant organisms may occur in combination of some genes, such as in high homozygosity. Sometimes they indicate the presence of certain specific actions breaks or simulates normal activity substance regulators—morphogens and hormones [9].

The relative magnitude of the FA is about 1% of the trait value [10]. Most approaches to the FA testing based on an evaluation of the variance differences between the left and right magnitudes in bilaterally symmetrical traits. The overcoming imperfection in methods of FA testing includes the standardizing of the size of the chosen traits, the sample size, and the number of traits selected for FA testing.

As a signal reflecting the environmental stress, FA is a typical attribute of the genetically determined norm of reaction of the organism. Two other types of asymmetry, the directional asymmetry and the antisymmetry, are implied as a genetic deviation from strict symmetry. The fluctuating asymmetry itself is related to nondirectional asymmetry. Some characteristics of three types of asymmetry are shown in **Table 1**.

The relation between the FA as a signal violating the DS and environmental stress of different nature, in spite of the large number of publications, is in the focus of scientific interest [11, 12].

According to Shmalgauzen's concept of stabilizing selection, the natural selection takes place even in a relatively stable and optimal ambient. It results in two

Sample characteristics	Fluctuating asymmetry	Directional asymmetry	Antisymmetry
R and L values	R = L	R ≠ L	R = L
Kurtosis, γ in descriptive statistics (R–L)	γ = 0–2	γ = 0–2	γ < 2
Type of frequency distribution (R–L)	Normal	Not normal	Normal/not normal

Note: R and L, values of right and left metric bilaterally symmetric traits; γ, mean of kurtosis.

Table 1.
Typical features of the three types of asymmetry.

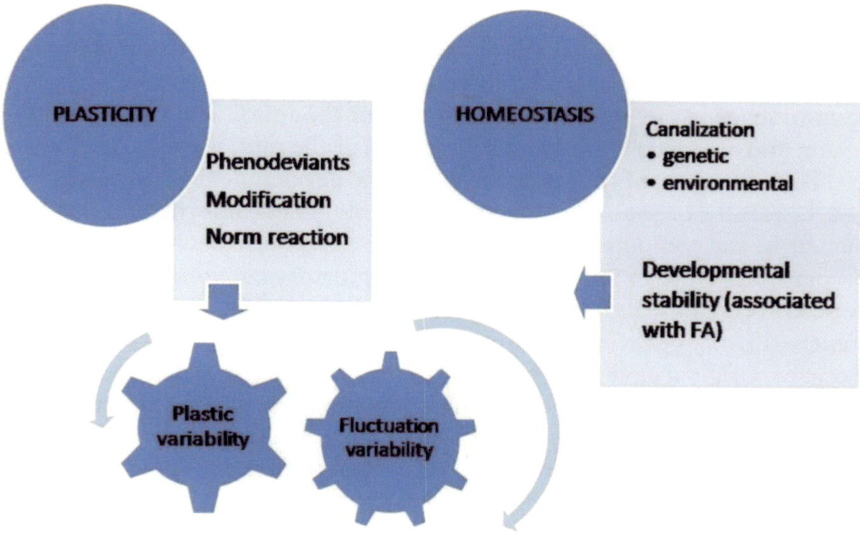

Figure 1.
Two types of phenotypic variability.

effects: in the elimination of mutants and recombinants and in the accumulation of mutations in a small population, which means the modification of the phenotype. By the term "canalization" Schmalhausen meant a tendency to achieve a certain genetically predetermined final state, different from the variability at the beginning of development [13]. Canalized selection addresses to the elimination of individuals that are susceptible to changing environmental factors. The term "ecological canalization," as an attribute of development homeostasis, has the meaning of stabilizing phenotypic variability [14]. The term "ecological plasticity" explains rather adaptive processes and characterizes the increased variability. Such a dialectical opposition pretends to be a source of microlevel adaptation in the population [15]. Developmental noise is the term that characterizes FA as an environmental origin phenomenon. On the other hand, the stability of development mostly is a genetic phenomenon which may vary in homo-/heterozygosity and depends on stress-reducing physiological status of the organism.

Thus, developmental stability of the organism is a biological characteristic depending on both causes as developmental noise (from the environment) and genetic variability characterizing the stages of ontogenesis.

Developmental homeostasis includes two attributes: developmental stability (FA is indicator) and environmental canalization. Plasticity and homeostasis lead to both types of phenotypic variability—plastic and fluctuation (**Figure 1**).

Phenotypic plasticity can help plants to overcome negative effects of some factors, first of all, temperature variability, and allow them to rapidly adjust traits to adverse conditions. The genetic variation could provide potential for adaptive evolution in response to changing climate variability [16].

2. Gene expression regulation

Expression of gene regulators of plant development is controlled by a number of internal and external factors. The internal factors affecting their activity include hormones, sucrose, and some mineral elements, and the external factors include temperature and light. In the regulation of differentiation and development, an important role is played by genes that contain promoters sensitive and specific to phytohormones and to such environmental factors as light and temperature. Currently, key genes have been identified that control embryogenesis, aging, and photomorphogenesis; regulate the functioning of the apical, lateral, and floral meristem; and are responsible for the formation of the root, leaves, and blood vessels. The expression of the genes regulating the development of flowers is best studied. Currently, based on the currently available genetic information, mathematical apparatus, and computer programs, key genes have been identified [17–19]. It has become possible to build so-called genetic regulatory networks (gene regulatory network, GRN), which allow to evaluate the whole range of interactions between different regulatory genes in the process of cell differentiation and the formation of plant organs. Original elements of these networks are capable of controlling several processes at different stages of development. Therefore, mutations affecting different parts of the same regulatory gene may differ in their phenotypic expression [20–22].

The MYB family (myeloblastosis) of transcription factors (TF) is known, which includes about 200 proteins that control such processes as root development, leaf patterning, trichome formation, cell cycle, circadian rhythms, and phytochrome signal transmission. A transcription factor (or sequence-specific DNA-binding factor) is a protein that controls the rate of transcription of genetic information from DNA to messenger RNA, by binding to a specific DNA sequence. The function of transcription factors is to regulate the on and off of genes for their expression in the right cell at the right time and in the right amount throughout cell and body life. Groups of TFs function in a coordinated fashion to direct cell division, cell growth, and cell death throughout life, cell migration, and organization during embryonic development and intermittently in response to signals from outside the cell, such as hormones [23].

A coactivator is a protein that works with transcription factors to increase the rate of gene transcription, whereas a corepressor is a protein that works with transcription factors to decrease the rate of gene transcription. Some of the transcription factors (AS1, AS2) are proteins responded for asymmetry in leaves.

Polycomb group proteins are important repressors of numerous genes in higher eukaryotes. In *Arabidopsis*, like heterochromatin protein 1 (LHP1), also known as TERMINAL FLOWER 2, was proposed as a subunit of polycomb repressive complex [24].

The theory of gene regulatory network dynamics can be represented in **Figure 2**:

The elementary units of the GRN are the genetic element G. It includes the pair (g, p), consisting of the g (gene) and the target (final) product synthesized p. The activity of the genetic element G is characterized by the rate of synthesis V of the target product p and the rate of its degradation/dissipation D. The target product

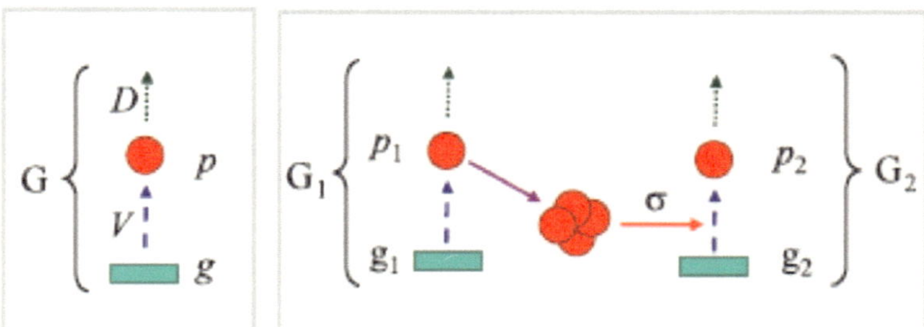

Figure 2.
Genetic elements and regulatory relationships of the GRN (on Kolchanov et al., 2013). G, G1, *and* G2 *are genetic elements, σ is the regulatory element,* V *is the synthesis of the target product, and* D *is the degradation/dissipation of the target product.*

can be either a RNA molecule or a protein. If a change in the concentration of the product *p1* decreases/increases the rate of change in the concentration of the product *p2*, there is a regulatory (oriented) relationship $\sigma = \sigma\ (G1,\ G2)$ between the genetic elements *G1* and *G2*.

3. Epigenetics as a new branch of genetics

3.1 Interspecific differences in the response to the environmental conditions

The different genotypes differed in their sensitivity to developmental noise. The heritable component of FA can thus be understood as a genetically modulated expression of variation that is itself entirely nongenetic [25, 26].

Epigenetics studies the stable changes in the level of gene expression that are not related to changes in the nucleotide sequence in DNA. For the first time, the term "epigenetics" was introduced by the embryologist and geneticist Conrad Waddington to describe changes in gene expression observed during the course of development. Epistatic interaction is reduced to the suppressor effect of one gene on another. Dominant epistasis explains the influence of the gene as the dominant allele.

The epigenetic status of the organism is determined by the nature and level of DNA methylation, posttranslational modifications of histones, the presence of histone isoforms, and the nature of chromatin stack in [27, 28]. The most well-described epigenetic mechanism of regulation is DNA methylation. Epigenetic variability is the variability of the population response rate, which leads to the realization of discrete states of morphological structures. In the era of postindustrial society, epigenetic variation becomes of particular interest. We can outline the next key aspects of experimental and field epigenetics studies. They are:

a. Frequency analysis of phenes as minor aberrations of morphogenesis

b. Spatial correlation of the level of phenetic differences between pairs of impact and control population on a continuous part of the areal

c. The use of methods of multidimensional ordination of phenetic compositions

d. Epigenetic landscape of a population

e. Comparison of dispersions of general asymmetry, fluctuating asymmetry, and directional asymmetry, characterizing manifestations of developmental destabilization at both individual and group levels

There are two genetic regulatory networks that play the main role in the manifestation of fluctuation and plastic variation. It is known that the heterogeneity of the sample value of metric or counting traits in a population, i.e., dispersion in heterogeneity, plays a key role in the magnitude of the revealed variability. The regulatory mechanisms of the epigenetic processes in the genome and at the cell level are the following: DNA methylation, histone code (posttranslational modifications of histones that occur by methylation, acetylation, phosphorylation, glycosylation, and ubiquity of histones), and followed by proteolysis [29, 30].

Genetic architecture and associations with fitness and with compensatory growth are studied under various stressful situations. Patterns in FAs play an important role in both DI and canalization of development. Their common origin and role in micro- and macroevolutionary processes are assumed [31, 32].

Phenogenetic variability is within the individual variability of the structure of antimers and metamers (homologue parts of plants). Asymmetry refers to the stochastic morphogenesis, i.e., random formation. The negative correlation between developmental instability and individual genetic diversity was obtained in flower petals within three *Opuntia echios* populations that suggest a buffer effect of heterozygosity on developmental instability [33].

The study showed that deletions do not affect the asymmetry of the *Drosophila* wing, but the development of canalization in a period of time changes. Consequently, genetic deletions have an impact on the canalization of the development of traits, but not on the stability of the development [34, 35]. Genetic assimilation is the subsequent genetic fixing of the new trait in the population. Some recent experimental and theoretical works have established a quantitative basis for these classic concepts of Waddington [36].

4. Plasticity and developmental stability

The genes responsible for plasticity, as well as environment and developmental stability, are closely related. Consider the next example. There are five metric traits in leaf blade of small-leaf linden *(Tilia cordata)*. When some traits of a leaf blade (measured distances between the veins) have a high plasticity, other ones expose instability of development with an increased FA value. FA is measured as $FA = L - R/ (L - R)$ in absolute value, where L and R are means of the left and right values of the bilaterally symmetrical trait. Plastic variability is measured as $PL = 1 - x/X$, where x and X are means of the smallest and largest values of trait size. The reason for the deviation was the ecological factors of the location of the populations and the climatic features of the vegetative season. As a whole, the plastic variability correlates to fluctuation variability. The more *PL* (i.e., trait size), the more FA. On the other hand, the correlation is small. Moreover there is an effect of conjugacy within one sample when the trait expressed the high index of plastic variability and another one expressed a high value of fluctuation variability (**Figure 3**).

The correlation is changeable and sometime can be negative. For example, in other sites, correlation r between *FA* in fourth trait and *PL* in fourth trait was -0.69 ($p < 0.05$; 2014).

Both types of variability, fluctuating and plastic, showed a conjugative effect. The correlative dependency deserves the focusing study. The level and character of expression of two type phenotypic variability depend on some factors. For plants the temperature presumably is a dominant factor. For example, the cold and humid vegetative season in 2017 played a main role.

The impact of climate especially noticeable among species. Two close species of linden (*Tilia cordata* and *Tilia platyphyllos*) were studied in one ecosystem. 2017 showed an increase in the plastic dimensions of leaf blades of the broad-leaved linden to the conditions of the Central Russian plain, which is attributed by the authors to one of the stress response phases, with increasing plastic variability of dimensional traits of leaf plates. We attribute this fact to the phenotypic deviation caused by low temperatures of air (**Figure 4**).

Increased fluctuating asymmetry was associated with the small size of the leaf blades. The growing plate, as is known, develops in pulsating mode, in the right/left side, and has a high asymmetry. Some signs, as we have seen, have shown a negative correlation between FA and Pl, and this relationship is of interest to phenogenetics [37].

In cultivated plants (wheat), a decrease in the stability of development occurred at elevated doses of fertilizer. It is assumed that this is a logical process, as a

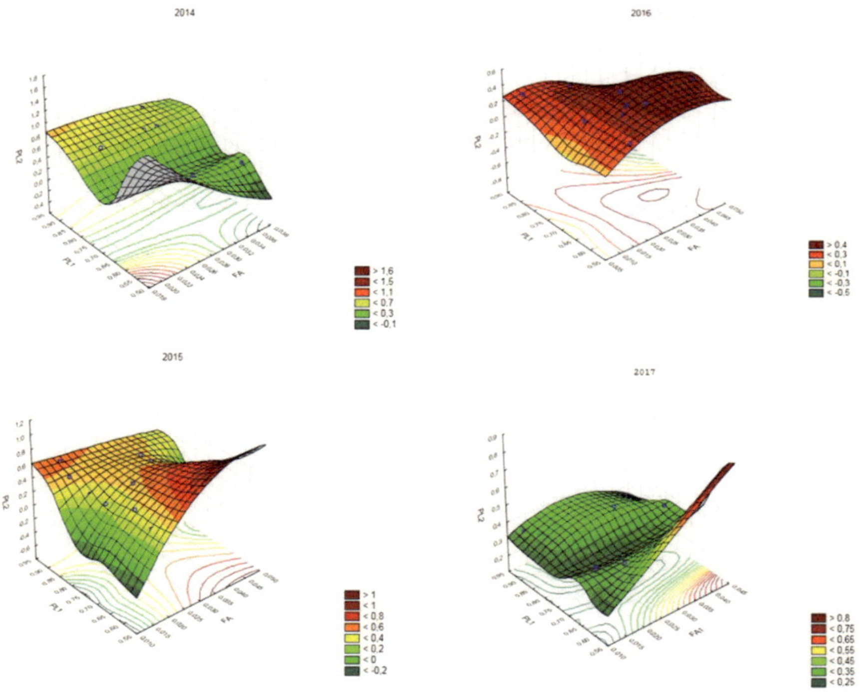

Figure 3.
Surface plots for correlation between the magnitude of plastic variability (PL, two traits, PL1 and PL2) and fluctuation variability (FA, one trait), Tilia cordata, *2014–2017, Moscow region, Russia (n = 100). 2016 year—Person's* r *= 0.73 (p < 0.05). Other year's* r *is not statistically significant.*

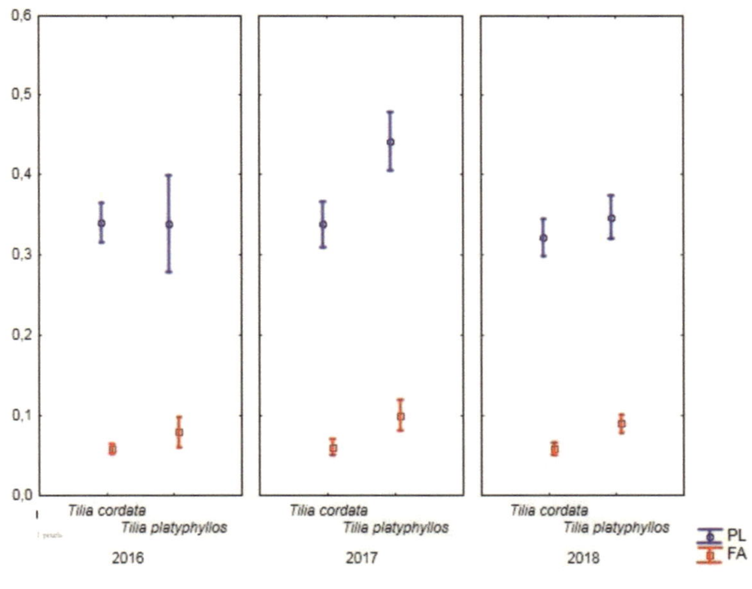

Figure 4.
Plot of means two types of variability PL (plasticity) and FA (fluctuating asymmetry) (both p < 0.05) and conf. Intervals (95%) for two species of Tilia *(2016–2018).*

response to high doses of fertilizer and excessive amplification of the vegetative and generative phases of plant development.

The gene regulatory network (AS1, AS2?) responds to temperature in the phase of the development of meristem. We conclude that stress factor of chemical pollution plays a secondary role. This confirms the fact of chemical load in site where fluctuating asymmetry was highest in 2017 and chemical pollution was less statistically significant. The mismatch between two types of variability in traits meant an absence of any correlation of FA-PL on some dimensional bilateral traits.

Proposed conclusion seems to be consensual, since in the literature on this issue there is a contradictory opinion on reciprocal effect of FA and PL [38, 39]. In other words, the traits were characterized by conjugation of two types of variability. The predominance of one type of variability is compensating by the weakness of other type of variability. For example, the weak fluctuation variability of trait No. 2 was compensated by its high plastic variability.

Thus the modularity of gene regulatory network associates with external factors. Based on the response to selection on the plasticity of a character, we concluded that plasticity is most likely due to epistasis. These models make some predictions with regard to correlations among trait plastic variety, developmental noise, and developmental instability. Modularity structure may enhance the adaptability and robustness of biological systems to perturbations [40, 41].

5. Methods of geometric morphometrics

The geometric morphometrics method allows to determine the value of the fluctuating asymmetry in combination of features and characteristics of the form of leaf blade and to test the presence directional asymmetry and antisymmetry. The degrees of freedom *df* were 14–18 times the number of degrees of freedom in the trivial two-factor analysis of variance with a high statistical significance ($p \leq 10^{-6}$) and an economy in the processing time and in the volume of the samples.

Methods of geometric morphometrics showed epistasis contributed ca. 20% of the variation in FA of size and 19% of the variation in FA of shape. This

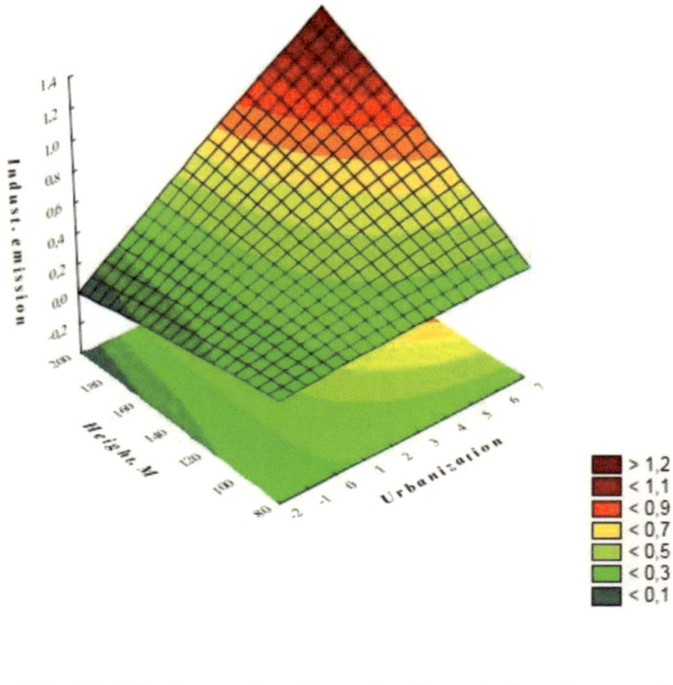

Figure 5.
Response surface of the desirability function depending on three factors (industrial emission, level of relief, and degree of urbanization) (n = 75).

Factor	*Acer platanoides*	*Tilia cordata*	*Betula pendula*	*Quercus robur*
Height	++++	+++++	+	+
Industrial emission	+++	+	++	+
Urbanization level	+	+	+	+
Vehicle emission	+++	++++++++++	+	++

Note: + − $p < 0.05$; ++ − $p < 0.01$ +++ − $p < 0.001$; ++++ − $p < 0.0001$; +++++ − $p < 0.00001$.

Table 2.
Factors influenced the developmental instability.

contribution was characteristic for the FA depending on the size of the studied traits of the mice [31]. Genetic screens in the model plant *Arabidopsis* have been particularly rewarding, identifying more than 130 epigenetic regulators. The diversity of epigenetic pathways in plants is remarkable, presumably contributing to the phenotypic plasticity of plant postembryonic development and the ability to survive and reproduce [42, 43]. Relief is an important factor determining plastic and fluctuation variability. The study of this factor using the example of birch *(Betula pendula)* showed that, together with other factors, such as the level of urbanization (on the graph in arbitrary units) and industrial emissions, an increase in the height of the relief reduces the stability of development (**Figure 5**).

The described interaction of three factors reflected a synergistic stress effect that influenced the stability of development. Differences in the range of reactions to the effect of each of the factors made it possible to compile a comparative scale for four plant species using the statistical significance p of the FA index, obtained by methods of geometric morphometrics (**Table 2**).

An autorepressor is a genetic element that provides, by a negative feedback mechanism, the regulation of expression of a gene encoding an autorepressor protein. Plastic and fluctuation variability are encoded by two genes. The mismatch between them occurs under stress, when regulatory proteins do not work on the principle of negative feedback. The regulatory gene network AS and AC2 seems to be a component of a wider regulatory network of plant development. Autorepressor appears as part of the GRN AS1 or AS2. Stress factors are simultaneously adaptive and destabilizing factors that temporarily disrupt the homeostasis of the development of the individual and the entire population.

6. Conclusion

This phenogenetic study contributes to the development of the epigenetic theory of gene networks. Bilateral asymmetry, reflecting phenotypic variation, is a convenient phenomenon for studying environmental dependencies in the frame of study of GRN and the direction of evolution.

Recently, there has been an increasing interest in system design information support for the storage and processing of phenotypic data and their integration with genomic information. To solve the problem of collection, integration, storage, and statistical processing of information about wheat plants, there are a number of computer systems, for example, WheatPGE (wheat-phenotype–genotype and environment).

The system stores various relationships describing the characteristics of an individual plant and allows you to uniquely establish the relationship between genotypic and phenotypic characteristics of plants as well as environmental parameters. The database is constantly being improved. Data on developmental stability and plate shape is one of the components of such a database. Application of the system will automate the receipt data on the relationship of genotype, phenotype, and environment in plants, including wheat, thereby contributing to the effective creation of new varieties of wheat with the economically important traits. As a whole the study of environmental factors affecting the manifestation of bilaterally symmetrical traits in plants is part of the research, in the framework of the study of the subtle mechanisms of GRN that are important for the ecology and evolution of plants.

Author details

Sergey Baranov[1,2*], Igor Vinokurov[1] and Lubov Fedorova[3]

1 Federal State Budgetary Scientific Institution "The Upper Volga Federal Agricultural Research Centre", Vladimir Region, Russia

2 Vladimir State University named after A.G. and N.G. Stoletov, Vladimir, Russia

3 Federal State Autonomous Educational Institution of Higher Education I.M. Sechenov First Moscow State Medical University of the Ministry of Health of the Russian Federation (Sechenov University), Russia

*Address all correspondence to: bar.serg58@gmail.com

References

[1] Waddington CH. Genetic assimilation of the bithorax phenotype Drosophila. Inform. Serv. 1954. p. 16

[2] Zaharov VM et al. Ontogenez i populjacija: ocenka stabil'nosti razvitija v prirodnyh populjacijah. Ontogenez. 2001;**32**(26):404-421

[3] Zaharov VM, Zjuganov VV. K ocenke asimmetrii bilateral'nyh priznakov kak populjacionnoj harakteristiki. Jekologija. 1980;**2**(1):10-16

[4] Van Dongen S, Molenberghs G, Matthysen E. The statistical analysis of fluctuating asymmetry: REML estimation of a mixed regression model. Journal of Evolutionary Biology. 1999;**12**:94-102

[5] Van Dongen S. Variation in measurement error in asymmetry studies: A new model, simulations and application symmetry. 2015;**7**:284-293. DOI: 10.3390/sym7020284

[6] Van Dongen S. Accuracy and power in the statistical analysis of fluctuating asymmetry: effects of between-individual heterogeneity in developmental instability. Ann. Zool. Fennici. 1999;**36**:45-52. Available from: https://www.jstor.org/stable/23735641

[7] Clarke GM. The genetic basis of developmental stability. IV. Individual and population asymmetry parameters. Heredity. 1998;**80**(5):553. DOI: 10.1046/j.1365-2540.1998.00326.x

[8] David M et al. A new measure of symmetry and its application to classification of bifurcating structures. Pattern Recognition. 2007;**40**(8): 2237-2250. DOI: 10.1016/j.patcog.2006.12.008

[9] Palmer AR, Strobeck C. In: Polak M, editor. Fluctuating Asymmetry Analysis Revisited I in Developmental Instability (DI): Causes and Consequences. Ist ed. New York: Oxford University Press; 2003. pp. 279-319

[10] Pape MA, Swaddle JP. Asymmetry, Developmental Stability and Evolution. UK: Oxford University Press; 1997

[11] Dongen SV. Fluctuating asymmetry and developmental instability in evolutionary biology: Past, present and future. Journal of Evolutionary Biology. 2006;**19**(6):1727-1743. DOI: 10.1111/j.1420-9101.2006.01175.x

[12] Shmal'gauzen II. Faktory jevoljucii (teorija stabilizirujushhego otbora). Vol. 445. Nauka: M.; 1968

[13] Debat V, David P. Mapping phenotypes: Canalization, plasticity and developmental stability. Trends in Ecology & Evolution. 2001;**16**(10): 555-561. Available from: http://tree.trends.com/

[14] Casal Jorge J et al. Signalling for developmental plasticity. Trends in Plant Science. 2004;**9**(6):309-314. DOI: 10.1016/j.tplants.2004.04.007

[15] Scheepens JF, Deng Y, Bossdorf O. Phenotypic plasticity in response to temperature fluctuations is genetically variable, and relates to climatic variability of origin, in *Arabidopsis thaliana*. AoB Plants. 2018. DOI: 10.1093/aobpla/ply043

[16] Chenlong L et al. The Arabidopsis SWI2/SNF2 chromatin remodeler BRAHMA regulates polycomb function during vegetative development and directly activates the flowering repressor gene SVP. PLoS Genetics. 2015;**11**(1):e1004944. DOI: 10.1105/tpc.15.00091

[17] Baum David A, Hileman LC. A developmental genetic model for the origin of the flower. Annual Plant

Reviews Online. 2018:3-27. Available from: https://onlinelibrary.wiley.com/doi/abs/10.1002/9780470988602

[18] de Wit M, Galvao VC, Fankhauser C. Light-mediated hormonal regulation of plant growth and development. Annual Review of Plant Biology. 2016;**67**:513-537. DOI: 10.1146/annurev-arplant-043015-112252

[19] Rik K et al. Epigenetic basis of morphological variation and phenotypic plasticity in *Arabidopsis thaliana*. The Plant Cell. 2015;**27**(2):337-348. DOI: 10.1105/tpc.114.133025

[20] Kyuha C et al. Regulation of microRNA-mediated developmental changes by the SWR1 chromatin remodeling complex. Plant Physiology. 2016;**171**:1128-1143. DOI: 10.1104/pp.16.00332

[21] Doroshkov AV et al. The evolution of gene regulatory networks controlling *Arabidopsis thaliana* L. trichome development. BMC Plant Biology. 2019;**19**:53. DOI: 10.1186/s12870-019-1640-2

[22] Mingli X et al. Developmental functions of miR156-regulated squamosa promoter binding protein-like (SPL) genes in *Arabidopsis thaliana*. PLoS Genetics. 2016;**12**(8):e1006263. DOI: 10.1371/journal.pgen.1006263

[23] Li Z, Li B, Liu J, et al. Transcription factors AS1 and AS2 interact with LHP1 to repress KNOX genes in Arabidopsis. Journal of Integrative Plant Biology. 2016;**58**:959-970. DOI: 10.1111/jipb.12485

[24] Peter KC, Barluenga M, Meyer A. Phenotypic plasticity, developmental instability, and robustness: The concepts and how they are connected. Frontiers in Ecology and Evolution. 2019;7:56. DOI: 10.3389/fevo.2019.00056

[25] Benedikt H et al. The developmental-genetics of canalization. Seminars in Cell & Developmental Biology. 2019;**88**

[26] Kumar S. Epigenetic memory of stress responses in plants. Journal of Phytochemistry & Biochemistry. 2018;**2**:1. DOI: 10.1016/j.semcdb.2018.05.019

[27] Jörn W, Hümpel A. Introduction to epigenetics. In: Epigenetics. Wiesbaden: Springer VS; 2017. pp. 11-29

[28] Peter R et al. Epigenetics and heterosis in crop plants. In: Epigenetics in Plants of Agronomic Importance: Fundamentals and Applications. Cham: Springer; 2019. pp. 129-147. DOI: 10.1007/978-3-319-07971-4_2

[29] Vladimir D, Dodd RS. Epigenetics: A potential mechanism for clonal plant success. Plant Ecology. 2015;**216**(2):227-233. Available from: http://scholar.google.ru/scholar?q=DOI+10.1007/s11258-014-0430-z&hl=en&as_sdt=0&as_vis=1&oi=scholart

[30] VanWallendael A et al. A molecular view of plant local adaptation: Incorporating stress-response networks. Annual Review of Plant Biology. 2019;**70**:559-583. DOI: 10.1146/annurev-arplant-050718-100114

[31] Mesa JM, Juvik JA, Paige KN. Individual and interactive effects of herbivory on plant fitness: Endopolyploidy as a driver of genetic variation in tolerance and resistance. Oecologia. 2019:1-10. DOI: 10.1007/s00442-019-04458-1

[32] Helsen P, Van Dongen S. Associations between floral asymmetry and individual genetic variability differ among three prickly pear (*Opuntia echios*) populations. Symmetry. 2016;**8**(11):116. DOI: 10.3390/sym8110116

[33] Lachowiec J, Queitsch C, Kliebenstein DJ. Molecular mechanisms

governing differential robustness of development and environmental responses in plants. Annals of Botany. 2015;**117**(5):795-809. DOI: 10.1093/aob/mcv151

[34] Takahashi KH, Daborn PJ, Hoffmann AA, Takano-Shimizu T. Environmental stress dependent effects of deletions encompassing Hsp70Ba on canalization and quantitative trait asymmetry in *Drosophila melanogaster*. PLoS One. 2011. DOI: 10.1371%2Fjournal.pone.0017295

[35] Spirov AV, Sabirov MA, Holloway DM. Systems Evolutionary Biology of Waddington's Canalization and Genetic Assimilation. In: Evolutionary Physiology and Biochemistry—Advances and Perspectives. IntechOpen; 2018. Available from: https://www.intechopen.com/books/perturbation-methods-with-applications-in-science-and-engineering/a-formal-perturbation-theory-of-carleman-operators

[36] Baranov SG, Zykov IE, Kuznetsova DD. Conjugacy of two types of phenotypic variability of small-leaved linden. Vavilov Journal of Genetics and Breeding. 2019;**23**: 496-502. DOI: 10.18699/VJ19.519

[37] Tonsor SJ, Elnaccash TW, Scheiner SM. Developmental instability is genetically correlated with phenotypic plasticity, constraining heritability, and fitness. Evolution. 2013;**67**(10): 2923-2935. DOI: 10.1111/evo.12175

[38] Tucić B et al. Phenotypic plasticity in response to environmental heterogeneity contributes to fluctuating asymmetry in plants: First empirical evidence. Journal of Evolutionary Biology. ;**2018, 31**(2):197-210. DOI: 10.5061/dryad.8th5m

[39] Klingenberg C. Analyzing fluctuating asymmetry with geometric morphometrics: Concepts, methods, and applications. Symmetry. 2015;**7**(2):843-934. DOI: 10.3390/sym7020843

[40] Long TA, Brady SM, Benfey PN. Systems approaches to identifying gene regulatory networks in plants. Annual Review of Cell and Developmental Biology. 2008;**24**:81-103. DOI: 10.1146/annurev.cellbio.24.110707.175408

[41] Pfluger J, Wagner D. Histone modifications and dynamic regulation of genome accessibility in plants. Current Opinion in Plant Biology. 2007;**10**(6):645-652. DOI: 10.1016/j.pbi.2007.07.013

[42] Klingenberg CP, Nijhout HF. Genetics of fluctuating asymmetry: A developmental model of developmental instability. Evolution. 1999;**53**(2): 358-375. DOI: 10.1111/j.1558-5646.1999.tb03772.x

[43] Baranov SG. Use of a geometric morphometric method to determine the developmental stability of *Betula pendula* Roth. Biology Bulletin. 2017;**44**(5):559-563. DOI: 10.1134/S1062359017050041

Section 4

Gene Evolution

Chapter 10

Sellafield, Seascale, and Scandinavia: A Legacy of Radioactive Contamination with Future Implications for Gene Evolution in Affected Ecosystems

Chanda Siddoo-Atwal

Abstract

Radioactive waste from nuclear installations and nuclear reprocessing plants, nuclear accidents, and radioactive fallout from nuclear weapons testing constitute a serious problem facing future generations. Marine algae and phytoplanktons accumulate radionuclides from their surroundings and are used as bioindicators of radioactive pollution in the environment. In Northern Europe, the affected marine systems include the Irish Sea, the Baltic Sea, and the North Sea. The main sources of this radioactive contamination are global fallout from nuclear weapons tests, river transport from Siberia, and marine transport of discharges from Sellafield and Chernobyl. An increased leukemia incidence has been observed in young children at Seascale near Sellafield, and an elevated incidence of leukemia has been recorded among young people (0–24 years) in the French canton of Beaumont-Hague close to the Cap de la Hague nuclear reprocessing facility. In Scandinavia, scientists suspect that people in parts of Sweden are still dying from cancer caused by radiation from the Chernobyl accident. Moreover, the Baltic Sea is contaminated with man-made plutonium radionuclides from nuclear reprocessing. However, some experts are able to dismiss the above relationships due to important uncertainties over the estimation of radiation doses from environmental discharges based on a mutational theory of carcinogenesis. Consequently, it appears to be of paramount importance to reevaluate the current methods for cancer risk assessment in the case of radiation exposure within the context of an apoptotic model of carcinogenesis that could explain such a discrepancy. According to this new model, subtle differences in gene expression in response to a carcinogen can initiate cell death or apoptosis and act as a trigger for carcinogenesis. Simultaneously, future implications for human gene evolution are unavoidable.

Keywords: nuclear reprocessing, plutonium, leukemia, lung cancer, alpha- and gamma-radiation carcinogenesis

1. Introduction

Sellafield (UK) and Cap de la Hague (France) are the two largest centers for commercial reprocessing of nuclear fuel in the world. This process involves the dissolution of spent fuel in boiling concentrated nitric acid, which results in the subsequent physicochemical separations of uranium and plutonium. Reprocessing operations release considerable volumes of liquid and gaseous wastes into the environment. These include large volumes of radioactivity, typically on a scale of several thousand more than that released by nuclear reactors.

1.1 Sellafield

In the case of Sellafield, aerial emissions have deposited two to three times the plutonium fallout from total atmospheric nuclear weapons testing within a 20-km radius of this facility. Since the inception of reprocessing at Sellafield in the 1960s, it has been estimated that between 250 and 500 kg of plutonium from the plant is now adsorbed onto sediments on the bed of the Irish Sea. Migration of these undersea deposits to coastal environments potentially represents a long-term hazard of unknown proportions [1].

The village of Seascale lies roughly south of Sellafield. A stretch of deserted beach runs northward toward the Sellafield plant, which has its discharge pipeline just 2 km off the coast. The beach used to be popular with bathers, but, now, many guide books make reference to the radioactive pollution from Sellafield. In fact, in 1993, a government survey found that the incidence of leukemia and non-Hodgkin's lymphoma was 14 times the national average and twice that in other areas of West Cumbria [2]. An increased incidence of retinoblastoma in children and a statistically significant increase in stillbirth risk in the Sellafield region have also been observed [1].

The main pathways for radiation exposure for people living in the vicinity of Sellafield are external radiation from airborne and deposited radionuclides, internal exposure resulting from inhalation of airborne radionuclides, and ingestion of radionuclides from contaminated food. In Sellafield, locally caught fish and shellfish have been found to be contaminated with toxic radionuclides, particularly, plutonium and americium [1].

1.2 Cap de la Hague

An elevated incidence of leukemia has been recorded among young people (0–24 years) in the French canton of Beaumont-Hague close to the Cap de la Hague nuclear reprocessing facility. In the latter case, two factors were found to be correlated strongly with increased leukemia risk: the use of local beaches for recreational activities by children and mothers during gestation and fish and shellfish consumption. In addition, a 3-year French epidemiological study in the region found a higher than expected morbidity in men from leukemia and respiratory cancers and leukemia and lung cancer in women [1].

1.3 Sea-to-land transfer

In Denmark, women have the highest rates of all forms of cancer in the Nordic countries [3], while their fish-eating neighbors in Norway have the highest rates of colorectal cancer for women in the world [4]. It is of some interest that Denmark, Ireland, and Norway have protested for many years over reprocessing at Sellafield and La Hague due to concern about the local impact on environment and health [1].

In parts of Sweden, scientists suspect that people are still dying from cancer caused by radiation from the Chernobyl accident [5]. Radioactive polonium, uranium, and plutonium have also been detected in the Baltic Sea where they are accumulated by various marine organisms. While the polonium and uranium isotopes may be derived from natural geological sources, the principle sources of plutonium in the Baltic Sea are radioactive fallout from nuclear weapons testing, releases from nuclear power plants and nuclear processing facilities at Sellafield and Cap de la Hague, and radioactive debris originating from Chernobyl [6].

It has now been established that a potentially important pathway for radioactive discharges to humans involves sea-to-land transfer since significant quantities of radionuclides can become airborne in seaspray and be transported inland by the wind [7]. Thus, concerns over nuclear reprocessing have led to a number of studies examining the health risks that may be associated with such nuclide discharges and, ultimately, have resulted in a scientific report commissioned by the European Parliament [1].

1.4 Scientific and technological options assessment

The findings of this report suggest that the Sellafield facility (closed in 2018) was found to be in violation of several key European requirements. Marine discharges at Sellafield led to significant radionuclide concentrations in certain foodstuffs exceeding EC limits (and, in this case, the Irish fishwives' tales about mutant fish were very likely true) [8]. Radionuclide discharges to the Sellafield marine environment resulted in doses to critical groups exceeding ten times the current UK and three times the EU limits. Increases of key radionuclide releases from in the late 1990s and projected future releases from Sellafield were not in keeping with European standards or their UK obligations as a member state of the European Union (EU). Generally, radionuclide concentrations in the Cap de la Hague environment have decreased since the 1980s, and calculated doses from routine releases remain well within EU limits. However, past accidents at La Hague have led to population doses significantly exceeding EU limits and are estimated to be responsible for 36% of the leukemia risk level for the 0–24-year age category around the site.

Contaminated marine environments can result in radiation exposure and radionuclide accumulation by marine life from plants to fish, and transfer to larger organisms such as mammals can occur via the food chain. Short-term health effects on humans can include various cancers, notably leukemia, caused by changes in gene regulation, and thyroid disorders. Long-term health effects may include heritable genetic mutations transmissible to future generations. This could have a potentially negative impact on the human gene pool and human evolution in affected areas, especially with the loss of individuals with the least mutations [9]. Ultimately, this may affect the fitness of the species and lead to a major decline or genetic degradation of select human populations [10].

2. Local health effects

2.1 Epidemiology

The main radioactive contaminants in the immediate vicinity of the Sellafield nuclear reprocessing facility have been identified as plutonium [Pu] and americium [Am] as a result of soil core samples taken from 95 locations within a few kilometers of the Sellafield nuclear complex. High levels of Pu-239 and Pu-240 have been deposited within a few hundred meters of the site, declining to lower levels within 3 km. The activity ratio of these radioisotopes indicates that the plutonium originated from uranium of low irradiation suggesting that deposition occurred during the early years of plant operation. The presence of Am-241 also correlates strongly with the Pu-239 and Pu-240 and is consistent with its derivation from Pu-241 deposition [11]. Pu-239 is formed by the spontaneous fission of U-238, which is a by-product of the uranium enrichment process routinely carried out at nuclear power plants [12].

In 1983, a cluster of cancers, most readily produced by ionizing radiation, were discovered in Seascale, a village situated 3 km from Sellafield, which is the principal nuclear reprocessing plant in the UK. It involved the occurrence of non-Hodgkin's lymphoma (NHL) with lymphoblastic leukemia in young people under the age of 25, who lived in Seascale in the period between 1955 and 1983 [13]. A subsequent scientific study confirmed that there was a marked excess of acute lymphatic leukemia (ALL) in Seascale [14]. In 1986, an excess of acute lymphatic leukemia was discovered near Dounreay, the site of the only other nuclear reprocessing plant in the UK [15]. Finally, in 1993, when an excess of acute lymphatic leukemia was also recorded in Egremont, a small town 7 km north of Sellafield, the emerging epidemiological pattern could no longer be ignored [14]. In fact, the incidence rates of ALL and NHL were increased to a similar extent in Seascale when the parents were incomers from other parts of the country or locals suggesting an environmental component to the problem [16]. Moreover, a significant excess of these cancers has continued to be observed in Seascale since this data was first collected.

Interestingly, one study measured the concentrations of Pu-239 and Pu-240 in the teeth of children throughout Great Britain and Ireland. Regression analysis showed that the concentrations of plutonium decreased with increasing distance from the Sellafield facility and suggests that this plant is not only a source of radioactive contamination locally, but it may affect a wider population within the British Isles [17]. An earlier study noted similar differences for plutonium levels in bone collected from people living in West Cumbria as compared with those living at sites remote from this location in the UK [18].

Further studies were commissioned by the Irish State to examine the possible effects of nuclear pollution from Sellafield on the coast of Ireland. The results showed a seacoast effect in coastal areas close to North Wales, particularly in towns where there was intertidal sediment contaminated with radioactive material from Sellafield or other historical sources. This was evidenced as a sharp rise in risk for leukemia in children and adults, especially near the northern entrance of the Menai Strait, which has fine intertidal sediment significantly contaminated with plutonium and other radioactive materials from Sellafield [19]. A follow-up localized these disturbing results in Caernarfon, Gwynedd, and Anglesey [20]. The explanation given was that sea-to-land transfer of radioactive particles followed by inhalation represented a risk to those living in the 0–1 km coastal strip since the particles could be transferred from the lungs to the lymphatic system resulting in leukemia. Such observations only serve to highlight the painful fact that acute uncertainties exist in the dosimetry of primary alpha-emitters, such as plutonium, in children and the fetus [21].

In autopsy tissues from west Cumbrian workers who had been employed in the nuclear energy industries, plutonium concentrations have been found to be generally higher than tissue concentrations in people from other regions of Great Britain. Furthermore, isotopic analysis using mass spectrometry has provided some evidence that this plutonium originated from aerial discharges from the British Nuclear Fuels Plant at Sellafield [22].

In radiation workers from the Sellafield plant, a significant excess of deaths has been recorded from cancer of the pleura and thyroid. In addition, a positive correlation has been observed between accumulated external radiation dose and mortality from leukemia, multiple myeloma, and all lymphatic and hematopoietic cancers.

There were also significant increases in risk with cumulative plutonium plus external radiation doses for all lymphatic and hematopoietic neoplasms. As a result, it has been concluded that the cancer incidence in Sellafield employees exposed to plutonium was significantly increased as compared with other radiation workers [23]. Furthermore, in a pooled cohort analysis of Sellafield (UK) and Mayak (Russia) workers, lung cancer risk from occupational plutonium exposure was studied. Poisson regression models provided a clear evidence of a linear association between cumulative internal plutonium lung dose and risk of lung cancer incidence and mortality in the pooled cohort [24].

2.2 Animal studies

Studies in animals have demonstrated that exposure to relatively large doses of plutonium (as compared with human doses) can cause tumors in the tissues in which it is retained [25].

As early as 1959, a single inhalation exposure to a smoke of plutonium oxide in rats was correlated with characteristic pathologic changes in the lungs at the sites where the material was deposited using autoradiographic and histopathologic methods. Moreover, the malignant tumors which occurred in many of the animals could be related to this severe focal damage in the lungs as a result of the plutonium oxide exposure [26]. It has also been recorded that a single intraperitoneal injection of monomeric Pu-239 in mice results in a significantly higher incidence of bone cancer in females than males, while castration of males equalizes the frequency of bone sarcomas in both sexes [27]. In another study, the cancer risk posed by Pu-239 was reduced in adult female mice by chelation therapy with subcutaneous injections of Zn Na_3 diethylenetriaminepentaacetate (Zn-DTPA) [28]. More recently, it was demonstrated that catechol-3, 6-bis(methyleiminodiacetic acid) (CBMIDA) is as effective as Zn-DTPA and Ca-DTPA in removing plutonium from the liver of rats and superior to both Zn-DTPA and Ca-DTPA in removing plutonium from bone [29].

Another significant study with rats found that point mutations in the p53 (or Tp53) tumor suppressor gene, which is involved in mediating apoptosis, seem to play a role in the development of lung tumors following inhalation exposure to plutonium dioxide [30]. However, these base transitions are not associated with nuclear accumulation of p53 protein suggesting that this may represent a later step in carcinogenesis involving resistance to apoptosis and cell transformation. Mutations in this gene are often encountered in lung tumors from uranium miners, and p53 appears to play a critical role in the cellular response to genetic damage caused by radiation in humans [31]. In addition, epigenetic inactivation of the p16 gene by methylation is common in rat lung tumors induced by Pu-239. The prevalence of p16 methylation in lung adenocarcinoma samples collected from Mayak plutonium workers is also significantly increased as compared with controls. The p16 protein has been reported to regulate apoptosis in diverse cell types [32].

The experimental inhalation of plutonium dioxide aerosols by beagle dogs revealed the long-term retention of plutonium in the lungs. Approximately 9% of the alveolar plutonium deposits were transferred to hilar lymph nodes by the first year and started to be detectable in abdominal lymph nodes about 2 years after inhalation [33]. More recent studies provide evidence that lung neoplasms can be induced in dogs by Pu-239 and can be associated with the expression of epidermal growth factor receptor [EGF-R] as in human lung tumors [34].

Furthermore, beagle dogs exposed to plutonium-nitrate and subjected to its retention in their tissues have been diagnosed with tumors in bronchioloalveolar, peripheral, and subpleural alveolar regions of the lung. The TUNEL assay revealed an elevation of apoptosis in the tracheal mucosa, tumor cells, and nuclear debris in the alveoli and lymph nodes of the beagles with statistically significant modifications in Fas ligand, B-cell lymphoma 2, and caspase-3 expression. It is of note that a comparably exposed human subject in the same study did not develop pulmonary tumors or display an elevated rate of apoptosis in lung tissues [35].

2.3 Cell culture studies

Cell studies on peripheral blood lymphocytes from workers with significant plutonium body burdens have revealed an increase in chromosome aberrations. Radiation-induced breakpoints were randomly distributed among the chromosomes according to length. However, the distribution of the breakpoints within the chromosomes displayed an excess in the centromeres and telomeres. This study suggests that plutonium depositions within the body can cause such aberrations since external radiation exposure was taken into account [36]. A further follow-up study performed on Sellafield workers with 20–50% and >50% maximum permissible body burdens (MPBB) of plutonium confirmed that these results are consistent with the hypothesis that hematopoietic precursor cells are being irradiated by internally deposited plutonium with subsequent selection resulting in only those cells with symmetrical aberrations reaching the peripheral lymphocyte pool [37]. Presumably, elimination of the nonviable cells occurs via programmed cell death or apoptotic pathways.

B lymphocyte precursor cells, which are the target cells for acute lymphoblastic leukemia in children, are highly susceptible to the lethal effects of α-particles and have a very low probability of surviving a single α-track [38]. Subsequently, alpha-irradiation has been reported to cause transmissible chromosomal instability, characterized by non-clonal cytogenetic aberrations with a high frequency of chromatid-type aberrations and a lower frequency of chromosome-type aberrations, in the clonal descendants of human hematopoietic stem cells. Delayed apoptotic cell death is evident in these clonal populations [39]. Furthermore, an interesting study demonstrated that both high-LET fast neutrons (of the type emitted by plutonium-239) and low-LET 60Co gamma rays induce apoptosis independently in resting human peripheral blood lymphocytes. Dose-response curves for each of these two types of radiation are characterized by an initial steep increase in the number of apoptotic cells below 1 Gy and a flattening of the curves at higher doses toward 5 Gy [40].

Apoptosis can also be selectively induced in transformed cells by neighboring normal cells via cytokine and reactive oxygen species (ROS)/reactive nitrogen species (RNS) signaling. In fact, in rat fibroblasts, high-LET α-particles are more effective than low-LET X-rays in stimulating intercellular induction of apoptosis for any given irradiated fraction of cells at very low doses. The increase in intercellular induction of apoptosis results from nitric oxide free radical (NO*) and peroxidase signaling mediated by transforming growth factor beta [TGF-β] [41]. In human fibroblasts following exposure to 0.1 and 1 Gy α-particles, irradiated populations display a dose-dependent increase in chromosome-type aberrations at the first cell division. Nonirradiated neighbor (bystander) fibroblast populations also demonstrate elevated chromatid-type aberrations. Both irradiated and bystander populations have significantly higher frequencies of chromatid aberrations over 25 doublings than controls [42]. These results support the increasing evidence that the biological effects of ionizing radiation are not limited to irradiated cells, but can extend to nonirradiated neighbors and be seen as genomic instability in their progeny at subsequent generations.

In another intriguing study, it was shown that human papillomavirus-immortalized human bronchial epithelial (BEP2D) cells can undergo malignant transformation when exposed to 1.5 Gy of α-particles emitted by a Pu-238 source [43]. A non-tumorigenic human thyroid epithelial cell line (HTori-3) has also been transformed into tumorigenic cells by exposure to alpha particles in vitro. These tumor cells were screened and found to contain experimentally induced mutations in the p53 tumor suppressor gene generated by irradiation of the human thyroid epithelial cell line [44].

3. Long distance ecosystem effects

3.1 Measurement of plutonium in sea water

Plutonium in the Baltic Sea has been assessed in precipitation, water, sediments, and microalgae. The major source of plutonium is fallout from nuclear tests, while European reprocessing facilities have contributed significant quantities of all plutonium isotopes. The Chernobyl accident contributed very little to overall plutonium concentrations of Pu-239 and Pu-240, whereas the Pu-238 and Pu-241 contributions were more significant [45]. Significantly increased concentrations of plutonium have also been found in the subsurface layers of the Norwegian and Greenland Seas. The plutonium activity ratio [^{238}Pu/^{239,240}Pu] found in both peaks displays a significant influence of Sellafield discharges [46]. Measurements of isotopic atomic ratios of plutonium in bottom deposit samples from the Norwegian and Greenland Seas show that the contribution of α activity of industrial plutonium from the European reprocessing plants, mainly Sellafield, on the background of global fallouts is 20–50% of total α activity (Pu-238-240) and 70–95% of total β activity (Pu-241) [47]. Moreover, a transfer factor for both conservative and non-conservative (including plutonium) Sellafield discharges has been detected in East Greenland Current Polar Water, a water mass which reflects contaminant levels in Arctic Ocean surface water [48].

3.2 Measurement of plutonium in plants

An interesting study analyzing Swedish lichen samples from various parts of Sweden demonstrated that in the areas most contaminated by the Chernobyl accident, the plutonium contributions from Chernobyl were as high as 85%. Sellafield-derived neptunium (Np) was found in brown seaweed samples from the west coast of Sweden 4–5 years after discharge (1995–1996), suggesting that it has greater mobility than plutonium in this particular marine environment [49]. Furthermore, the long-lived anthropogenic radionuclides Np-237, Pu-239, and Pu-240 were determined in marine environmental samples (seaweed and seawater) from Swedish-Danish waters and the North Atlantic Ocean at various locations. Most of the Np-237 in these waters is considered to originate from the Sellafield plant with some contribution from global fallout. There was quite a wide variation in Np-237 activity concentrations in *Fucus vesiculosus* and Pu-240/Pu-239 atomic ratios in *F. vesiculosus* samples [50]. In Norway, *F. vesiculosus* samples collected in the North Sea, Skagerrak, and in selected fjords also contain Pu-239 and Pu-240 concentrations with relatively large fluctuations from year to year with a slowly decreasing trend in the activity concentrations from 1980 to 2010, possibly coinciding with declining Cap de la Hague emissions. However, seawater samples collected in the North Sea near Scotland do not display a similar decreasing trend in Pu-239 and Pu-240 radioactivity, thereby suggesting a Sellafield effect [51].

3.3 Measurement of plutonium in marine biota

Plutonium-239 and plutonium-240 concentrations in Baltic biota have revealed that these radionuclides are strongly accumulated by some species such as algae, benthic animals, and fish [52]. In the southern Baltic Sea, the Pu-239 and Pu-240 concentrations in surface seawater samples range from 5.2 mBq m^{-3} for Gdańsk Bay to 150 mBq m^{-3} for Pomeranian Bay, and plutonium has been found to be distributed in parts of cod gills and skin [53]. At the same time, it is important to note that the major fraction of plutonium in the Baltic Sea is rapidly associated with sediments and only 1% is present in the water column. Therefore, in view of the above findings, it is of the utmost importance to identify safe scavenging processes and filtration techniques for future remediation efforts [45]. Some of the remediation strategies previously adopted for surface waters affected by the Chernobyl accident include adding chemicals to bind radionuclides, construction of traps to contain contaminated sediments, and dredging of contaminated deposits [54].

4. Possible mechanisms of alpha- and gamma-radiation carcinogenesis

4.1 An apoptotic model of carcinogenesis

Classically, experimental carcinogenesis is a complex, multistage process including initiation, promotion, and malignant progression in which the failure of DNA repair mechanisms and the subsequent clonal expansion of damaged cells play a pivotal role. However, more recently, it has become apparent that the pathogenesis of cancer is closely connected with aberrantly regulated apoptotic cell death and the resulting deregulation of cell proliferation [55, 56].

The Ames assay as a universal test for carcinogenicity was based on the classical model of carcinogenesis involving the failure of DNA repair mechanisms and the subsequent clonal expansion of mutated cells. However, mutagenicity in bacterial strains is not always an indicator of carcinogenicity since many carcinogens are not mutagenic [57]. Although this may be one feasible mechanism of carcinogenesis in laboratory models, it does not adequately fit many existing systems of carcinogenesis which are increasingly connected with the dysregulation of apoptotic mechanisms in the cell.

Somatic cell and germline point mutations do occur, and the frequency of mutation can be accelerated by external factors like mutagenic substances such as ionizing radiation. Spontaneously, this is still considered to be a relatively slow process and, probabilistically, more likely to have an effect on a later step of carcinogenesis involving cell immortalization or transformation than to initiate carcinogenesis, or, more likely, even to affect future generations. According to Darwinian doctrine, random mutations in the general population result in the gradual evolution of a species. With exposure to mutagens in the environment, though, this genetic evolution can be sped up resulting in the acquisition of new traits or characteristics by a particular group. Ultimately, there is always the possibility that this unnaturally accelerated evolution may negatively impact the fitness of the species.

In the two-stage model of tumor formation (**Figure 1**), step I exposure to a carcinogen, possibly facilitated by a genetic predisposition, results in an epigenetic or genetic event causing continuous apoptotic activation of cells in the target tissue. In step II, when the carcinogen may or may not be present, resistance to apoptosis and continuous cell proliferation result due to another genetic or epigenetic event.

The extended proliferation of these somatic cells can result in the development of the neoplastic phenotype [58]. According to this model based on human skin carcinogenesis, while an epigenetic change is more likely to occur in the first step of carcinogenesis, a genetic mutation is more likely to occur in the second step of carcinogenesis as a function of probability [59]. A nice example is provided by the p53 mutations which appear to inhibit cell death in mammalian lung tumors induced by plutonium dioxide.

4.2 Alpha- and gamma-radiation-induced apoptosis

It is well known that radioactive metals and their isotopes routinely have long half-lives and many emit gamma radiation. As an example, the half-life of uranium-238 [U-238] is 4.46 billion years, while that of U-235 is 704 million years, and, although they are primarily alpha-emitters, both also emit gamma radiation. In addition, some plutonium radioisotopes with intermediate half-lives (up to thousands of years) emit gamma rays along with alpha and beta particles [60]. Thus, the potential health effects on wildlife and humans are virtually unlimited in terms

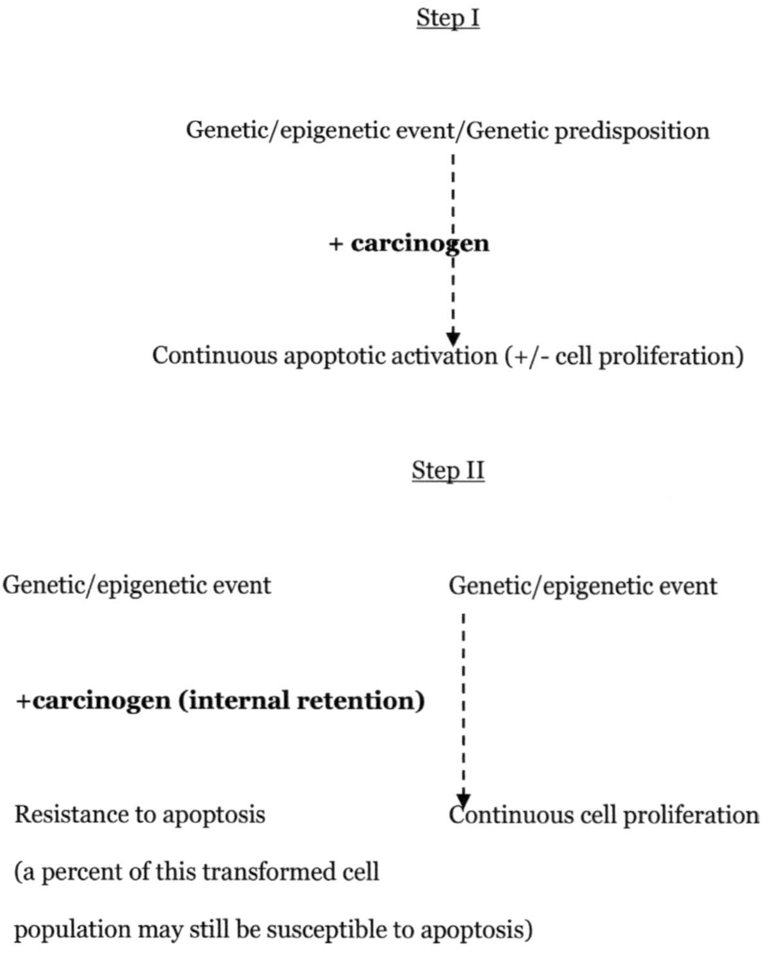

Figure 1.
Two-stage model of tumor formation. Step I exposure to a carcinogen, possibly facilitated by a genetic predisposition, results in an epigenetic or genetic event causing continuous apoptotic activation of cells in the target tissue. In step II, when the carcinogen may or may not be present, resistance to apoptosis and continuous cell proliferation result due to another genetic or epigenetic event.

of duration once the environment becomes contaminated with significant levels of such radioisotopes due to nuclear accidents or nuclear reprocessing activities.

Alpha-irradiation has been reported to cause delayed apoptotic cell death in the clonal descendants of human hematopoietic stem cells. Another study has demonstrated that high-LET fast neutrons (of the type emitted by plutonium-239) induce apoptosis in resting human peripheral blood lymphocytes. Apoptosis can also be selectively induced in transformed mammalian cells by neighboring normal cells in response to high-LET α-particles via cytokine and reactive oxygen species (ROS)/ reactive nitrogen species (RNS) signaling.

Like UV-rays, ϒ-rays are considered to be a complete carcinogen. In one large study, 400 days (a little over 1 year) of continuous external ϒ-ray exposure induced tumors in a significant percentage of treated mice (>86.7%) and resulted in their premature deaths [61]. External exposure to ϒ-rays has been reported to stimulate apoptosis in certain cell types including rat thymocytes, mouse myeloid cells, and human lymphoid cells. Lymphoblastoid cell lines derived from healthy individuals undergo p53-independent apoptosis in response to high-LET radiation and p53-dependent apoptosis in response to low-LET radiation [62]. Furthermore, some human carcinoma cell lines exposed to internal ϒ-radiation via the decay of a DNA-incorporated radionuclide [I-125] display radiosensitivity as a result of activation of apoptotic pathways [63]. Thus, exposure to both external and internal ϒ-radiation sources can stimulate apoptotic cell death.

4.3 Uranium and apoptotic pathways

It appears that different isotopic compositions of uranium like many other heavy metals can penetrate to the subcellular level resulting in bioaccumulation and initiation of oxidative stress in zebrafish tissues [64]. At subtoxic concentrations (>100 μM), depleted uranium precipitates mainly in the nucleus of the human kidney, liver, and neuronal cell lines [65]. Chemically, uranium can activate oxygen species in the course of redox reactions via the redox chemistry of transition metals [66] and enhance free radical production via the ionization phenomenon induced by alpha particle emissions to produce DNA damage in the form of double-strand breaks [67, 68]. Although uranium can emit alpha, beta, and gamma radiation, alpha particle emissions are of the greatest relevance in relation to depleted uranium, which is a waste product of uranium enrichment [69]. Thus, uranium is capable of initiating chemotoxicity and radiotoxicity [70] via the generation of free oxygen species and possibly via more direct mechanisms involving DNA damage in cells. Both biochemical pathways can stimulate cell death or apoptosis, which has been linked to carcinogenesis in various cancer models. In fact, the loss of equilibrium between cell death and cell proliferation in a tissue may play a critical role in tumor formation (**Figure 1**) [71, 72].

4.4 Plutonium and apoptotic pathways

Plutonium is mainly deposited in the human lungs, liver, and skeleton, where it is retained for many years. Pu-239 has a physical half-life of 24, 110 years and a biological half-life of 20–50 years [73]. Intravascular administration of plutonium citrate in the human body results in radioactive concentration in the bone marrow, the liver, the bone, the spleen, the kidney, and the lungs [74]. Environmental transfer is determined by its solubility in human body fluids such as its measurement via urine bioassays [25]. This environmental plutonium transfer can give rise to α-particle radiation doses of ≈3–7 μSv/a in human bone and ≈10–20 μSv/a in the liver [12], which may result in chemo- and radiotoxicity via the ionization phenomenon induced by alpha particle emissions similar to that caused by depleted uranium. Previously, the incidence of multiple primary cancers in Nagasaki plutonium atomic bomb survivors has been associated with radiation exposure [75]. Interestingly, recent autoradiographic analysis of paraffin-embedded tissue samples from Nagasaki atomic bomb victims (who died within 5 months after the bombing) has confirmed the presence of alpha-tracks from deposited plutonium, and, so, the contribution of internal plutonium radiation exposure [Pu-239] to overall toxicity cannot be dismissed [76]. Thus, a two-stage model of tumor formation in the case of plutonium is plausible, possibly with a degree of apoptosis occurring in the transformed cell population due to continued exposure to the internalized plutonium (**Figure 2**).

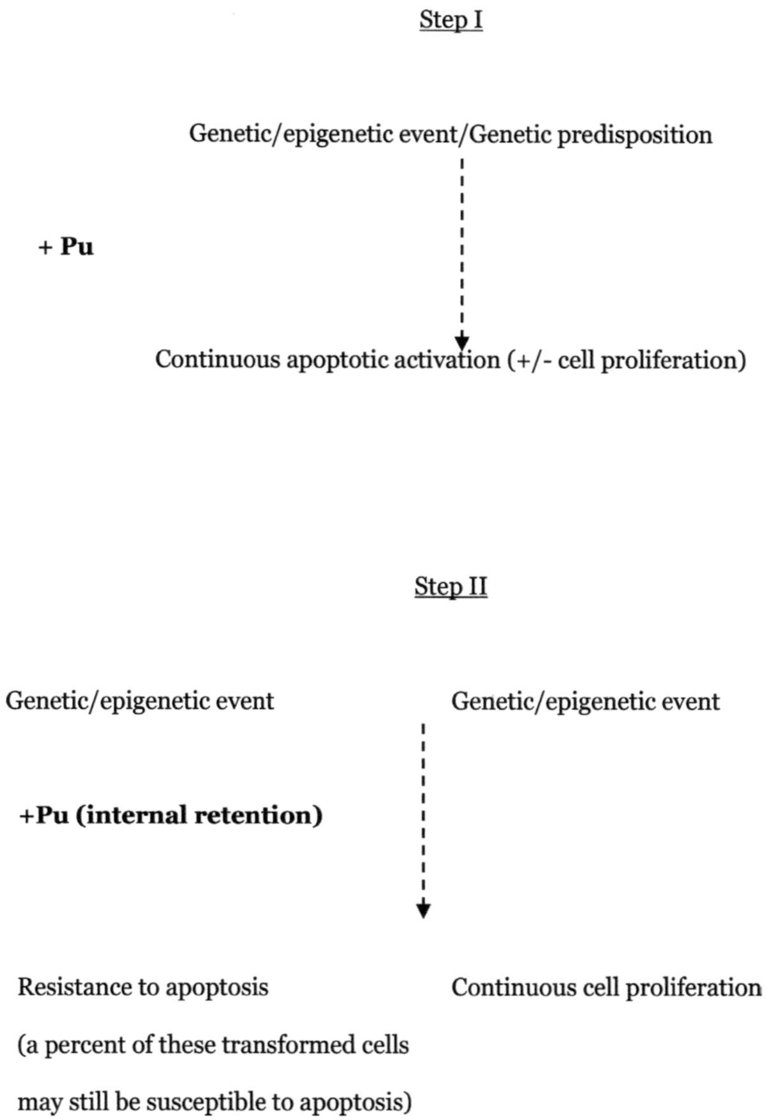

Figure 2.
Two-stage model of tumor formation for plutonium. A two-stage model of tumor formation in the case of plutonium is plausible, possibly with a degree of apoptosis occurring in the transformed cell population due to continued exposure to the internalized plutonium.

5. Discussion

Epidemiological studies in the north of England revealed a cluster of cancers most readily produced by ionizing radiation in three villages situated close to nuclear reprocessing plants. When an elevated incidence of acute lymphatic leukemia was observed in children living in all three locations, the emerging pattern could no longer be ignored. Subsequently, the deposition of radioactive plutonium was found in the immediate vicinity of the Sellafield plant, and locally caught fish and shellfish were found to be contaminated with toxic radionuclides, particularly, plutonium. A study that measured the concentration of plutonium in the teeth of children in Great Britain and Ireland concluded that it decreased with increasing distance from the Sellafield facility and suggested that this plant was not only a source of radioactive contamination locally, but that it may be affecting a wider population within the British Isles. Further studies commissioned by the Irish State found a sharp rise in the risk for leukemia in children and adults, especially near the northern entrance of the Menai Strait, which has fine intertidal sediment significantly contaminated with plutonium and other radioactive materials from Sellafield due to the sea-to-land transfer of radioactive particles. Plutonium concentrations in west Cumbrian workers from the nuclear energy industries were generally higher at autopsy than tissue concentrations in people from other regions of Great Britain, and isotopic analysis confirmed that this plutonium originated from aerial discharges from the British Nuclear Fuels Plant at Sellafield. In fact, the cancer incidence in Sellafield employees exposed to plutonium was significantly increased as compared with other radiation workers. A positive correlation was observed between accumulated external radiation dose and mortality from leukemia, multiple myeloma, and all lymphatic and hematopoietic cancers. There were also significant increases in risk with cumulative plutonium plus external radiation doses for all lymphatic and hematopoietic neoplasms. Moreover, a combined study between Sellafield and Mayak nuclear reprocessing plants found evidence of a linear association between cumulative internal plutonium lung dose and risk of lung cancer incidence.

Animal studies in rats revealed that a single inhalation of plutonium oxide smoke can result in malignant tumors of the lungs. A single intraperitoneal injection of plutonium-239 in mice particularly predisposes females to bone cancer. Plutonium dioxide inhalation in rats is associated with lung tumors displaying point mutations in the p53 tumor suppressor gene. Mutations in p53 are found in lung tumors from uranium miners, as well, suggesting a similar mechanism of carcinogenesis. Inactivation of the p16 gene by methylation is common in rat lung tumors induced by Pu-239. Methylation of p16 is also observed in lung adenocarcinomas from Mayak plutonium workers. Lung neoplasms can be induced in beagle dogs by Pu-239 and can be associated with the expression of epidermal growth factor receptor [EGF-R] as in human lung tumors. Furthermore, beagle dogs exposed to plutonium-nitrate have been diagnosed with tumors in bronchioloalveolar, peripheral, and subpleural alveolar regions of the lung.

Chromosome studies with peripheral blood lymphocytes from plutonium workers may provide indirect evidence that hematopoietic precursor cells are being irradiated by internally deposited plutonium with subsequent selection and elimination of badly damaged cells via apoptosis. B lymphocyte precursor cells are highly susceptible to the lethal effects of α-particles, and high-LET fast neutrons (of the type emitted by plutonium-239) induce apoptosis in resting human peripheral blood lymphocytes. Delayed apoptotic cell death is evident in clonal descendants of human hematopoietic stem cells exposed to alpha-irradiation. Apoptosis can also be selectively induced in transformed rat fibroblasts by neighboring normal cells via cytokine and reactive oxygen species (ROS)/reactive nitrogen species (RNS) signaling. It has been shown that human papillomavirus-immortalized human bronchial epithelial (BEP2D) cells can undergo malignant transformation when exposed to 1.5 Gy of α-particles emitted by a Pu-238 source.

6. Conclusions

Thus, according to a new method for cancer risk assessment based on an apoptotic model of carcinogenesis, there is sufficient evidence to support that plutonium is a carcinogen associated with the etiology of leukemia and lung cancer in humans, especially when it is internalized from the environment. Comparative apoptosis studies in normal bronchial epithelial cells exposed independently to α-particles from a Pu-239 source and plutonium citrate might be useful in exploring the differing effects of external and internal alpha-irradiation.

In addition, recent results from many laboratories show that chromosomal instability in proliferating mammalian cells is characterized by a persistent increase of chromosomal aberrations and rearrangements occurring de novo during successive cell generations. This phenotype can be induced equally by low- and high-LET irradiation in lymphocytes at very low doses of exposure, conceivably contributing to cancer risk [42, 77]. The potential future implications in select human groups living in environments contaminated with radioactive substances are staggering and may include ailing populations with the elimination of fit individuals from the gene pool due to persistent illness or disease.

Moreover, the possibility of detecting the biological effects of Sellafield radioactive contamination as far afield as Scandinavia is a distinct reality since significant quantities of radionuclides can become airborne in seaspray and be transported inland by the wind for long distances. Plutonium from the Sellafield plant has been identified in coastal areas of North Wales where it has been correlated with an increased risk for leukemia. It has also been detected in the Baltic Sea, the Norwegian and Greenland Seas, and in seawater and seaweed collected from Swedish-Danish waters. Furthermore, plutonium-239 and plutonium-240 have been found in the gills and skin of cod from the Baltic Sea. Therefore, combined radiation risks including a Sellafield contribution may be elevating cancer rates in Scandinavian populations.

In this day and age of increasingly widespread radioactive pollution, heavy metal chelation therapy is an essential addition to all modern hospitals. As an example, injections of a chelation agent, Zn-DTPA, have been found to be useful for plutonium detoxification and reducing the plutonium-associated cancer risk in female mice. Although the biochemical pathways for plutonium detoxification in the human body are not well understood, free radicals are eliminated via various antioxidant activities including enzymes such as superoxide dismutase, catalase, glutathione peroxidase, and glutathione S-transferase. Zinc supplementation has also been reported to be potentially beneficial in ameliorating the health hazards of nuclear waste such as depleted uranium, which may generate free radicals as a result of similar cellular mechanisms to plutonium. Thus, zinc absorption via the regular use of zinc oxide-based sunscreens may provide some protection against certain heavy metal environmental pollutants [78, 79].

Finally, this grave environmental issue begs the question of moral culpability and social responsibility. Therefore, it is suggested that radionuclide remediation [79] should be practiced and promoted by violators of international nuclear safety regulations, Sellafield in this particular case. Compensation should be meted out to the injured parties in the case of nuclear accidents at Sellafield and Cap de la Hague. Finally, the frequency of accidents and future costs of such negligence at nuclear and reprocessing installations including the possible decimation of the species should be weighed carefully by countries when considering the attractive attributes of alternative clean energy sources like wind, water, and solar power. Once nuclear power plants start springing up around the world and producing vast quantities of radioactive waste, there will be no way of turning back the clock and decontaminating the planet.

Author details

Chanda Siddoo-Atwal
Moondust Cosmetics Ltd., West Vancouver, Canada

*Address all correspondence to: moondustcosmetics@gmail.com

References

[1] Schneider M, Coeytaux X, Faïd YB, Marignac Y, Rouy E. Possible toxic effects from the nuclear reprocessing plants at Sellafield and Cap de la Hague. European Parliament. STOA Scientific and Technological Options Assessment. 2001:1-168

[2] 1999-2019, Core, Cumbrians Opposed to a Radioactive Environment. Available from: http://corecumbria.co.uk

[3] Denmark still worst country in the Nordics for cancer. The Local. 2017. Available from: www.thelocal.dk

[4] Bazilchuk N. Cancer mystery in Norway. ScienceNordic. 2017. Available from: http://sciencenordic.com

[5] Abdelrahman R. Swedes still dying from Chernobyl radiation. The Local. 2007. Available from: www.thelocal.se

[6] Skwarzec B. Polonium, uranium and plutonium in the Southern Baltic Sea. AMBIO. 1997;**26**(2):113-117

[7] McKay WA, Pattenden NJ. The transfer of radionuclides from sea to land via the air a review. Journal of Environmental Radioactivity. 1990;**12**:49-77

[8] Martin J. Ireland could sue British operators over contamination. The Irish Post. 2014. Available from: www.irishpost.com

[9] Muller HJ. The relation of recombination to mutational advance. Mutation Research/Fundamental and Molecular Mechanisms of Mutagenesis. 1964;**1**(1):2-9. DOI: 10.1016/0027-5107(64)90047-8

[10] Loewe L, Hill WG. The population genetics of mutations: Good, bad and indifferent. Philosophical transactions of the Royal Society of London. Series B, Biological Sciences. 2010;**365**(1544):1153-1167. DOI: 10.1098/rstb.2009.0317

[11] Jones SR, Willans SM, Smith AD, Cawse PA, Baker SJ. Deposition of actinides in the vicinity of Sellafield, Cumbria: Accounting for historical discharges to atmosphere from the plant. Science of the Total Environment. 1996;**183**(3):213-229. DOI: 10.1016/0048-9697(96)05067-X

[12] Taylor DM. Environmental plutonium in humans. Applied Radiation and Isotopes. 1995;**46**(11):1245-1252. DOI: 10.1016/0969-8043(95)00167-C

[13] Doll R. The Seascale cluster: A probable explanation. British Journal of Cancer. 1999;**81**(1):3-5

[14] Craft AW, Parker L, Openshaw S, Charlton M, Newell J, Birch JM, et al. Cancer in young people in the north of England, 1968-85: Analysis by census wards. Journal of Epidemiology and Community Health. 1993;**47**(2):109-115

[15] Heasman MA, Kemp IW, Urquhart JD, Black R. Childhood leukemia in Northern Scotland. Lancet. 1986;**1**:266

[16] Dickinson H, Parker L. Quantifying the effect of population mixing on childhood leukaemia risk: The Seascale cluster. British Journal of Cancer. 1999;**81**(1):144-151

[17] O'Donnell RG, Mitchell PJ, Priest ND, Strange L, Fox A, Henshaw DL, et al. Variations in the concentration of plutonium, strontium-90 and total alpha-emitters in human teeth collected within the British Isles. Science of the Total Environment. 1997;**201**(3):235-243

[18] Popplewell DS, Ham GJ, Johnson TE, Barry SF. Plutonium in

autopsy tissues in Great Britain. Health Physics. 1985;**49**(2):304-309

[19] Busby C. Radiation from Sellafield and Cancer near the Irish Sea: Second Report from the Irish Sea Group in Support of the Litigation Short and Others vs. BNFL and the Attorney General. Aberystwyth: Green Audit; 2000

[20] Busby C. Nuclear Pollution, Childhood Leukaemia, Retinoblastoma and Brain Tumours in Gwynedd and Anglesey Wards near the Menai Straits, North Wales 2000-2003. Green Audit; 2004

[21] Crouch D. Science and trans-science in radiation risk assessment: Child cancer around the nuclear fuel reprocessing plant at Sellafield, U.K. Science of the Total Environment. 1986;**53**(3):201-216. DOI: 10.1016/0048-9697(86)90133-6

[22] Popplewell DS, Ham GJ, McCarthy W, Morgan M. Isotopic composition of plutonium in human tissue samples determined by mass spectrometry. Radiation Protection Dosimetry. 1989;**26**(1-4):313-316. DOI: 10.1093/oxfordjournals.rpd.a080423

[23] Omar RZ, Barber JA, Smith PG. Cancer mortality and morbidity among plutonium workers at the Sellafield plant of British nuclear fuels. British Journal of Cancer. 1999;**79**:1288-1301

[24] Gillies M, Kuznetsova I, Sokolnikov M, Haylock R, O'Hagan J, Tsareva Y, et al. Lung cancer risk from plutonium: A pooled analysis of the Mayak and Sellafield Worker Cohorts. Radiation Research. 2017;**188**(6):725-740. DOI: 10.1667/RR14719.1

[25] Clarke RH, Dunster J, Nenot J-C, Smith H, Voeltz G. The environmental safety and health implications of plutonium. Journal of Radiological Protection. 1996;**16**(2):91

[26] Lisco H. Autoradiographic and histopathologic studies in radiation carcinogenesis of the lung. Laboratory Investigation. 1959;**8**(1)

[27] Taylor GN, Gardner P, Mays CW, Wrenn ME, Charrier K. Incidence of plutonium-induced bone cancer in neutered mice. Cancer Research. 1981;**41**(3):971-973

[28] Jones CW, Mays CW, Taylor GN, Lloyd RD, Parker SM. Reducing the risk of Pu-239 by chelation therapy. Radiation Research. 1986;**107**(3):296-306. DOI: 10.2307/3576834

[29] Fukuda S, Iida H, Hseih YY, Chen W. Effects of CBMIDA [Catechol-3, 6-bis (methyleiminodiacetic acid)] on removal of plutonium in rats. Japanese Journal of Health. 1992;**27**(1):11-15. DOI: 10.5453/jhps.27.11

[30] Yamada Y, Oghiso Y. Mutations in Tp53 gene sequences from lung tumors in rats that inhaled plutonium dioxide. Radiation Research. 1999;**152**(6s):S107-S109. DOI: 10.2307/3580125

[31] Kelly G, Stegelmeier BL, Hahn FF. p53 alterations in plutonium-induced F344 rat lung tumors. Radiation Research. 1995;**142**(3):263-269. DOI: 10.2307/3579134

[32] Belinsky SA, Kling DM, Liechty KC, March TH, Kang T, Gilliland FD, et al. Plutonium targets the p16 gene for inactivation by promoter hypermethylation in human lung adenocarcinoma. Carcinogenesis. 2004;**25**(6):1063-1067. DOI: 10.1093/carcin/bgh096

[33] Bair WJ, Thompson RC. Plutonium: Biomedical research. Science. 1974;**183**:715-722

[34] Gillett NA, Stegelmeier BL, Kelly G, Haley PJ, Hahn FF. Expression

of epidermal growth factor receptor in plutonium-239-induced lung neoplasms in dogs. Veterinary Pathology. 1992;**29**:46-52. DOI: 10.1177/030098589202900106

[35] Nielsen CE, Wang X, Robinson RJ, Brooks AL, Lovaglio J, Patton KM, et al. Carcinogenic and inflammatory effects of plutonium-nitrate retention in an exposed nuclear worker and beagle dogs. International Journal of Radiation Biology. 2014;**90**(1):60-70. DOI: 10.3109/09553002.2014.859765

[36] Tawn EJ, Hall JW, Schofeld GB. Chromosome studies in plutonium workers. International Journal of Radiation biology and Related Studies in Physics, Chemistry and Medicine. 1985;**47**(5):599-610. DOI: 10.1080/09553008514550831

[37] Whitehouse CA, Tawn EJ, Riddell AE. Chromosome aberrations in radiation workers with internal deposits of plutonium. Radiation Research. 1998;**150**(4):459-468. DOI: 10.2307/3579666

[38] Griffiths SD, Marsden SJ, Wright EG, Greaves MF, Godhead DT. Lethality and mutagenesis of B lymphocyte progenitor cells following exposure to α-particles and X-rays. International Journal of Radiation Biology. 1994;**66**(2):197-205. DOI: 10.1080/09553009414551101

[39] Kadhim MA, Lorimore SA, Townsend KMS, Goodhead DT, Buvkle VJ, Wright EG. Radiation-induced genomic instability: Delayed cytogenetic aberrations and apoptosis in primary human bone marrow cells. International Journal of Radiation Biology. 1995;**67**(3):287-293. DOI: 10.1080/09553009514550341

[40] Vral A, Cornelissen M, Thierens H, Louagie H, Philippe J, Struckmans K, et al. Apoptosis induced by fast neutrons versus 60Co gamma-rays in human peripheral blood lymphocytes. International Journal of Radiation Biology. 2009;**73**(3):289-295. DOI: 10.1080/095530098142383

[41] Abdelrazzak AB, Stevens DL, Bauer G, O'Neill P, Hill MA. The role of radiation quality in the stimulation of intercellular induction of apoptosis in transformed cells at very low doses. Radiation Research. 2011;**176**(3):346-355. DOI: 10.1667/RR2509.1

[42] Ponnaiya B, Jenkins-Baker G, Bigelow A, Marino S, Geard CR. Detection of chromosomal instability in α-irradiated and bystander human fibroblasts. Mutation Research/Fundamental and Molecular Mechanisms of Mutagenesis. 2004;**568**(1):41-48. DOI: 10.1016/j.mrfmmm.2004.06.045

[43] Sun J-F, Sui J-L, Zhou P-K, Geng Y, Hu Y-C, Cao Z-S. Decreased efficiency of γ-ray-induced DNA double-strand break rejoining in malignant transformants of human bronchial epithelial cells generated by alpha-particle exposure. International Journal of Radiation Biology. 2002;**78**(9):773-780. DOI: 10.1080/09553000210141441

[44] Gamble SC, Myfanwy CC, Riches AC, Herceg Z, Bryant PE, Arranda JE. p53 mutations in tumors derived from irradiated human thyroid epithelial cells. Mutation Research/Fundamental and Molecular Mechanisms of Mutagenesis. 1999;**425**(2):231-238. DOI: 10.1016/S0027-5107(99)00045-7

[45] Holm E. Plutonium in the Baltic Sea. Applied Radiation and Isotopes. 1995;**46**(11):1225-1229. DOI: 10.1016/0969-8043(95)00164-9

[46] Herrmann J, Nies H, Goroncy I. Plutonium in the deep layers of the Norwegian and Greenland Sea. Radiation Protection Dosimetry.

1998;**75**(1-4):237-245. DOI: 10.1093/oxfordjournals.rpd.a032237

[47] Stepanov AV, Tsvetkov OS, Tishkov VP, Belyaev BN, Donkin VD, Ivanova LM, et al. Isotopic composition of plutonium in the bottom deposits of the Norwegian Sea and Greenland Sea and identification of the sources of contaminationhearder. Atomic Energy. 1999;**87**(4):745-752

[48] Dahlgaard H. Transfer of European coastal pollution to the arctic: Radioactive tracers. Marine Pollution Bulletin. 1995;**31**(1-3):3-7. DOI: 10.1016/0025-326X(95)00003-6

[49] Lindahl P. studies of long-lived radionuclides in the environment—With emphasis on ^{99}tc, ^{237}np and pu isotopes [thesis dissertation]. Lund University Hospital, Department of Radiation Physics; 2003. ISBN: 91-974444-0-5

[50] Lindahl P, Roos P, Holm E, Dahlgaard H. Studies of Np and Pu in the marine environment of Swedish–Danish waters and the North Atlantic Ocean. Journal of Environmental Radioactivity. 2005;**82**(3):285-301. DOI: 10.1016/j.jenvrad.2005.01.011

[51] Gafvert T, Heldal H, Brungot L, Gwynn J, Stralberg E, Svaeren I, Stromsnes H, Kolstad A, Moller B, Komperod M, Lind B, Rudjord A. Radioactivity in the Marine Environment 2010. Results from the Norwegian Marine Monitoring Programme (RAME). StrålevernRapport. 2012

[52] Skwarzec B. Polonium, uranium and plutonium in the southern Baltic ecosystem. Czechoslovak Journal of Physics. 1999;**49**(1):461-466

[53] Strumińska DI, Skwarzec B. Plutonium concentrations in waters from the southern Baltic Sea and their distribution in cod (*Gadus morhua*) skin and gills. Journal of Environmental Radioactivity. 2004;**72**(3):355-361. DOI: 10.1016/S0265-931X(03)00220-0

[54] Yao Y, Volchek K, Lambert P, Brown CE. Environmental Impacts of the Fukushima and Chernobyl Accidents and their Remediation: A Review. Ottawa, Canada: Emergencies Science and Technology, Environment Canada, Atomic Energy of Canada Limited, International Safety Research Inc.; 2014

[55] Denmeade SR, Isaacs JT. Programmed cell death (apoptosis) and cancer chemotherapy. Cancer Control. 1996;**3**(4):303-309

[56] Martin KR. Targeting apoptosis with dietary bioactive agents. Experimental Biology and Medicine. 2006;**231**:117-129

[57] Lijinsky W. Chemistry and Biology of N-Nitroso Compounds. The Edinburgh Building, Cambridge, UK: Cambridge University Press; 1992

[58] Mondello C, Chiodi I. Cellular immortalization and neoplastic transformation simultaneous, sequential or independent? Telomeres, telomerase or karyotypic variations? Cell Cycle. 2013;**12**(11):1804-1805

[59] Siddoo-Atwal C. Heavy metal carcinogenesis: A possible mechanistic role for apoptosis. Vegetos. 2017;**30**(Special):125-132. DOI: 10.5958/2229-4473.2017.00046.5

[60] USNRC—Backgrounder on plutonium. 2017. Available from: www.nrc.gov

[61] Tanaka IB, Tanaka S, Ichinohe K, Matsushita S, Matsumoto T, Otsu H, et al. Cause of death and neoplasia in mice continuously exposed to very low dose rates of gamma rays. Radiation Research. 2007;**167**(4):417-437

[62] Meijer AE, Ekedahl J, Castro JB, et al. High-LET radiation induces

apoptosis in lymphoblastoid cell lines derived from ataxia-telangiectasia patients. International Journal of Radiation Biology. 2001;**77**(3):309-317

[63] Urashima T, Nagasawa H, Kassis A, et al. Induction of apoptosis in human tumour cells after exposure to augur electrons: Comparison with gamma-ray exposure. Nuclear Medicine and Biology. 2006;**33**(8):1055-1063

[64] Barillet S, Adam C, Palluel O, Devaux A. Bioaccumulation, oxidative stress, and neurotoxicity in *Danio rerio* exposed to different isotopic compositions of uranium. Environmental Toxicology & Chemistry. 2007;**26**(3):497-505

[65] Rouas C, Bensoussan H, Suhard D, Tessier C, Grandcolas L, Rebiere F, et al. Distribution of soluble uranium in the nuclear cell compartment at subtoxic concentrations. Chemical Research in Toxicology. 2010;**23**(12):1883-1889

[66] Yazzie M, Gamble SL, Civitello ER, Stearns DM. Uranyl acetate causes DNA single strand breaks in vitro in the presence of ascorbate (vitamin C). Chemical Research in Toxicology. 2003;**16**(4):524-530

[67] Miller AC, Stewart M, Brooks K, Shi L. Page N depleted uranium-catalyzed oxidative DNA damage: Absence of significant alpha particle decay. Journal of Inorganic Biochemistry. 2002;**91**(1):246-252

[68] Chapter 5: Potential human health effects of uranium mining, processing, and reclamation. In: Uranium Mining in Virginia: Scientific, Technical, Environmental, Human Health and Safety, and Regulatory Aspects of Uranium Mining and Processing in Virginia. National Academies Press; 2012

[69] Hon Z, Osterreicher J, Navratil L. Depleted uranium and its effects on humans. Sustainability. 2015;**7**(4):4063-4077

[70] Ng CY, Pereir S, Cheng SH, Adam-Guillermin C, Garnier-Laplace J, Yu KN. Combined effects of depleted uranium and ionizing radiation on zebrafish embryos. Radiation Protection Dosimetry. 2015;**167**(1-3):311-315

[71] Siddoo-Atwal C. AT, apoptosis, and cancer: A viewpoint. Indian Journal of Ecology. 2009;**36**(2):103-110

[72] Siddoo-Atwal C. Chapter 7—An approach to cancer risk assessment and carcinogenic potential for three classes of agricultural pesticides. In: Peshin R, Dhawan AK, editors. Natural Resource Management: Ecological Perspectives. Springer; 2019

[73] Human Health Fact Sheet—Plutonium. 2001. Available from: www.globalsecurity.org

[74] Langham WH, Bassett SH, Harris PS, Carter RE. Distribution and excretion of plutonium administered intravenously to man. Health Physics. 1980;**38**:1031-1060

[75] Nakashima M, Kondo H, Miura S, Soda M, Hayashi T, Matsuo T, et al. Incidence of multiple primary cancers in Nagasaki atomic bomb survivors: Association with radiation exposure. Cancer Science. 2007;**99**:87-92

[76] Shichijo K, Takatsuji T, Fukumoto M, Nakashima M, Matsuyama MM, Sekine I. Autoradiographic analysis of internal plutonium radiation exposure in Nagasaki atomic bomb victims. Heliyon. 2018;**4**(6). DOI: 10.1016/j.heliyon.2018.e00666

[77] Lambert B, Holmberg K, Hackman P, Wennborg A. Radiation induced chromosomal instability in human T-lymphocytes. Mutation Research/Fundamental and Molecular Mechanisms of Mutagenesis.

1998;**405**(2):161-170. DOI: 10.1016/S0027-5107(98)00133-X

[78] Siddoo-Atwal C. A case-study of uranium and heavy metal detoxification. Indian Journal of Ecology. 2019;**46**(2):427-430

[79] Siddoo-Atwal C. Chapter 17—Biological effects of uranium and its decay products on soil microbes, plants, and humans. In: Varma A et al., editors. Plant Microbe Interface. Springer; 2019. (In press)